Fluid Catalytic Cracking Handbook

Fluid Catalytic Cracking Handbook

Editor

Gaurav Vashishth

scitus
academics

Fluid Catalytic Cracking Handbook
Edited by **Gaurav Vashishth**

Printed in 2017

ISBN: 978-1-68117-373-3

Library of Congress Control Number: 2015941561

© 2016 by
SCITUS Academics LLC,
616, Corporate Way, Suite 2, 4766,
Valley Cottage, NY 10989

www.scitusacademics.com

Notice

Contents

Preface

This is providing practical information and tools that engineers can use to maximize the profitability and reliability of their fluid catalytic cracking operations. The updated chapters and new content deliver expertise and know-how to an industry that faces significant cost cutting in capital expenditure and R&D, along with the retirement of technical specialists who are taking existing knowledge out of the industry with them. This FCC Handbook provides a valuable easy-to-understand resource for both experienced and inexperienced engineers and anyone else associated with the FCC process. Fluid catalytic cracking (FCC) is one of the most important conversion processes used in petroleum refineries. It is widely used to convert the high-boiling, high-molecular weight hydrocarbon fractions of petroleum crude oils to more valuable gasoline, olefinic gases, and other products.[1][2][3] Cracking of petroleum hydrocarbons was originally done by thermal cracking, which has been almost completely replaced by catalytic cracking because it produces more gasoline with a higher octane rating. It also produces byproduct gases that are more olefinic, and hence more valuable, than those produced by thermal cracking.

Editor

Analysis of Process Variables via CFD to Evaluate the Performance of a FCC Riser

H. C. Alvarez-Castro[1], E. M. Matos[1], M. Mori[1], W. Martignoni[2], and R. Ocone[3]

[1]School of Chemical Engineering, University of Campinas, 500 Albert Einstein Avenida, 13083-970 Campinas, SP, Brazil

[2]PETROBRAS/AB-RE/TR/OT, 65 República do Chile Avenida, 20031-912 Rio de Janeiro, RJ, Brazil

[3]Chemical Engineering, Heriot-Watt University, Edinburgh EH144AS, UK

ABSTRACT

Feedstock conversion and yield products are studied through a 3D model simulating the main reactor of the fluid catalytic cracking (FCC) process. Computational fluid dynamic (CFD) is used with Eulerian-Eulerian approach to predict the fluid catalytic cracking behavior. The model considers 12 lumps with catalyst deactivation by coke and poisoning by alkaline nitrides and polycyclic aromatic adsorption to estimate the kinetic behavior which, starting from a given feedstock, produces several cracking products. Different feedstock compositions are considered. The model is compared with sampling data at industrial operation conditions. The simulation model is able to represent accurately the products behavior for the different operating conditions considered. All the conditions considered were solved using a solver ANSYS CFX 14.0. The different operation process variables and hydrodynamic effects of the industrial riser of a fluid catalytic cracking (FCC) are evaluated. Predictions from the model are shown and comparison with experimental conversion and yields products are presented; recommendations are drawn to establish the conditions to obtain higher product yields in the industrial process.

INTRODUCTION

Since the first FCC commercial riser, many improvements have been achieved which have helped the process reliability and its capacity to transform heavier feedstock at relatively low costs; currently the FCC process remains the primary conversion process in the petrochemical industry. For a number of refineries, the fluid catalytic cracking remains the main source of profitability and the accomplishment of its operation decides the market competiveness of the cracking unit. Approximately 350 FCC units are in operation worldwide, with over 12.7 million barrels per day as total capacity. Most of the existing cracker units have been designed or modified by six major technology licensers [1]. The design of each FCC unit can be different but their common target is to upgrade low-cost hydrocarbons to more valuable products. FCC and ancillary units, such as the alkylation unit, are responsible for about 45% of the gasoline produced worldwide. Papers have flourished in recent years in the attempt to describe and simulate numerically the phenomena observed in such process. To predict the solid and gas phase behavior the Eulerian-Eulerian approach has been used due to low computational effort required [2]. In this study, the Eulerian-Eulerian approach is used, where the solid phase is treated as a continuum [3–5]. Computational fluid dynamic (CFD) was implemented to solve discretized equations; a hybrid mesh (tetrahedral mesh with refining prisms at the wall) was used as the calculation grid. The 12-lump kinetic model proposed by Wu et al. [6] with catalyst deactivation was coupled with the hydrodynamic model to evaluate the full problem. The lumping approach has been studied to describe the kinetic behavior of catalytic cracking with a large number of components where each lump is constituted by hundreds of kinds of molecules in a specific range of molecular weights. The methodology is shown to be very powerful when a large number of components are involved [7–11]. The simulation model uses a 12-lump effect that the variation of different process variables has on the conversion network which has the advantage of representing with good reliability the products and presents the option of representing the feedstock through three different lumps. The purpose of this study is to predict the yield and conversion behaviors at different operating conditions in the industrial riser of a FCC unit, with a 12-lump kinetic model. Different operational conditions have been studied, in order to estimate product yields.

RISER PROCESS

The riser is the main equipment of the FCC unit. Inside the riser the feedstock is fed through nozzles and mixture with the catalyst and the accelerant steam in the injection zone. The performance of the nozzles to guarantee fast vaporization of the feedstock and a good contact of the gasoil droplets with the catalyst is key to improve the FCC riser efficiency; the feedstock nozzles are positioned about

5–12 meters above the bottom of the reactor. In accordance with kind of FCC design, the number of feedstock injections can be from 1 to 15. Practically all of the riser reactions take place between 1 and 3 s. Reactions start as soon as the feed enters in contact with the hot catalyst.

The increasing velocity due to the vapor production acts as the means to carry the catalyst up in the riser. The hot solid supplies the necessary heat to vaporize the feedstock and bring its temperature to the temperature needed for cracking, compensating, also, for the reducing in temperature due to endothermic behavior of riser reactions. Standard risers are designed for an outlet velocity of 12–18 m/s. During the operation, coke deposits on the catalyst, declining the catalyst activity and thus representing a concern for the efficiency of the cracking reactions [12].

MATHEMATICAL MODEL

The fluid dynamic equations and kinetic model are summarized in Section 3.1 and taken and adapted from Alvarez-Castro [13]; the catalytic cracking kinetic models are taken from Wu et al. [6] and Chang et al. [14]. In order to study the heterogeneous, kinetics, and the particle phase deactivation, (15)–(20) were implemented in the CFX code.

Governing Equations for Transient Two Fiuid Models

Governing Equations

(1)Gas-solid fluid model (Eulerian-Eulerian) [15]:

$$\frac{\partial}{\partial t}\left(\varepsilon_g \rho_g\right) + \nabla \cdot \left(\varepsilon_g \rho_g \mathbf{u}_g\right) = 0,$$

$$\frac{\partial}{\partial t}\left(\varepsilon_s \rho_s\right) + \nabla \cdot \left(\varepsilon_s \rho_s \mathbf{u}_s\right) = 0,$$

(1)

where ε is the volume fraction, ρ is density, and $\mathbf{u}$ is the velocity for each phase.

(2) Momentum equations:

$$\frac{\partial}{\partial t}\left(\varepsilon_g \rho_g \mathbf{u}_g\right) + \nabla \cdot \left(\varepsilon_g \rho_g \mathbf{u}_g \mathbf{u}_g\right)$$

$$= \nabla \cdot \left[\varepsilon_g \mu_g \left(\nabla \mathbf{u}_g + \left(\nabla \mathbf{u}_g\right)^T\right)\right] + \varepsilon_g \rho_g \mathbf{g} - \varepsilon_g \nabla p + M,$$

$$\frac{\partial}{\partial t}\left(\varepsilon_g \rho_g \mathbf{u}_s\right) + \nabla \cdot \left(\varepsilon_s \rho_s \mathbf{u}_s \mathbf{u}_s\right)$$

$$= \nabla \cdot \left[\varepsilon_s \mu_s \left(\nabla \mathbf{u}_s + \left(\nabla \mathbf{u}_s\right)^T\right)\right] + \varepsilon_s \rho_s \mathbf{g} - \varepsilon_s G \nabla \varepsilon_s - M,$$

(2)

where p is the pressure, μ the viscosity, G the modulus of elasticity, g the acceleration of gravity, and M the interphase momentum transfer:

$$M = \left(150\frac{\varepsilon_s^2 \mu_g}{\varepsilon_g d_s^2} + \frac{7}{4}\frac{|\mathbf{u}_s - \mathbf{u}_g| \varepsilon_s \rho_g}{d_s}\right)\left(\mathbf{u}_s - \mathbf{u}_g\right)$$

(3)

For dense zones where, $\varepsilon_s > 0.2,$

$$M = \left(\frac{3}{4}C_d \frac{|\mathbf{u}_s - \mathbf{u}_g| \varepsilon_s \varepsilon_g \rho_g}{d_s}\right)\left(\mathbf{u}_s - \mathbf{u}_g\right)$$

For dilute zones where, $\varepsilon_s < 0.2,$

where d_s is the solid diameter and C_d is the drag coefficient

$$C_d = \frac{0.44}{\varepsilon_g^{2.65}}$$

(Re > 1000,

adequate for inertial effects to govern viscous effects),

$$C_d = \frac{1}{\varepsilon_g^{2.65}} \frac{24}{\text{Re}} \left(1 + 0.15 \text{Re}^{0.687}\right)$$

(Re < 1000, viscous and inertial effects are significant),

$$G = \exp\left[C_G \left(\varepsilon_s - \varepsilon_{s,\max}\right)\right] \tag{4}$$

(see [16]), where $\varepsilon_{s,\max}$ is maximum volume fraction and the packing limited about 0.65.

(3) Turbulence equations:

$$\mu_g = \mu_{\text{lam},g} + \mu_{\text{tur},g} \quad \text{(effective viscosity)} \tag{5}$$

(a) The -epsilon mixture model [17]:

$$\mu_{\text{tur},g} = C_\mu \rho_g \frac{k^2}{\epsilon}, \tag{6}$$

where k is the turbulence kinetic energy, ϵ is the turbulence eddy dissipation, and C_μ is constant

$$\frac{\partial}{\partial t}\left(\rho_g k\right) + \nabla \cdot \left(\rho_g \mathbf{u}_g k\right) = \nabla \cdot \left[\mu_{\text{lam},g} + \frac{\mu_{\text{tur},g}}{\sigma_k} + \nabla k\right]$$

$$+ P^k - \rho_g \epsilon,$$

$$\frac{\partial}{\partial t}\left(\rho_g \epsilon\right) + \nabla \cdot \left(\rho_g \mathbf{u}_g \epsilon\right) = \nabla \cdot \left[\mu_{\text{lam},g} + \frac{\mu_{\text{tur},g}}{\sigma_\epsilon} + \nabla \epsilon\right]$$

$$+ \frac{\epsilon}{k}\left(C_{\epsilon,1} P^k - C_{\epsilon,2} \rho_g \epsilon\right), \tag{7}$$

where σ_k, σ_ϵ, $C_{\epsilon,1}$, and $C_{\epsilon,2}$ are constants. P^k is the turbulence production

$$P^k = \mu_{\text{tur},g}\left(\nabla u_g + \left(\nabla u_g\right)^T\right) \tag{8}$$

(4)Heat transfer model:

$$\frac{\partial}{\partial t}\left(\varepsilon_g \rho_g H_g\right) + \nabla \cdot \left(\varepsilon_g \rho_g u_g H_g\right)$$

$$= \nabla \cdot \left(\varepsilon_g \lambda_g \nabla T_g\right) + \gamma \left(T_s - T_g\right)$$

$$+ \varepsilon_g \rho_g \sum_r \nabla H_r \frac{\delta C_r}{\delta t} - Q_R - Q_V, \tag{9}$$

$$\frac{\partial}{\partial t}\left(\varepsilon_s \rho_s H_s\right) + \nabla \cdot \left(\varepsilon_s \rho_s u_s H_s\right) = \nabla \cdot \left(\varepsilon_s \lambda_s \nabla T_s\right) + \gamma \left(T_g - T_s\right) \tag{10}$$

where H is enthalpy, T temperature, λ thermal conductivity, Q_R heat of cracking reactions, and Q_V energy lost in gasoil vaporization

$$\gamma = \frac{\mathrm{Nu}\lambda}{d_s}, \tag{11}$$

where γ is the interphase heat transfer coefficient, d_s is the diameter of the catalyst, and Nu is the Nusselt number

$$\mathrm{Nu} = 2 + 0.6\sqrt{\mathrm{Re}}\,\mathrm{Pr}^{0.3} \tag{12}$$

(see [18]).

(5)Energy lost in gasoil vaporization transfer by hot solid:

$$Q_V = 400\,\mathrm{kJ/kg\ of\ gasoil} \tag{13}$$

(see [5]).

(6)Kinetic model [6].

Variation of the chemical species:

$$\frac{\partial}{\partial t}\left(\varepsilon_g \rho_g C_{g,I}\right) + \nabla \cdot \left(\varepsilon_g \rho_g u_g C_{g,I}\right) = \nabla \cdot \left(\varepsilon_g \Gamma_i \nabla C_{g,I}\right) + \hat{R}_I, \tag{14}$$

where Γ is diffusivity, $C_{g,I}$ is the concentration of species I in the gas phase, and $\hat{R}_I$ is consumption or formation of each species.

(6.1)The rate equation for the generic reaction:

$$\hat{R}_{I,r} = -k_r \cdot \rho_p \cdot (\rho\alpha_i) \cdot \phi(t) \cdot F(N) \cdot F(A), \quad (15)$$

where we have the following: $\phi(t)$, catalyst poisoning due to coke content, $F(N)$, alkaline nitrides, $F(A)$, polycyclic aromatic adsorption, k_r, kinetic constant, ρ_p, particle density, and $(\rho\alpha_i)$, the mass content of species i in gaseous phase.

(a)Decay model based on coke content:

$$\Phi(t) = e^{(-\alpha t)}, \quad (16)$$

where we have (t), time, and α, constant.

(b)Alkaline nitrides:

$$F(N) = \frac{1}{1 + k_N C_N t_C / F_{c/o}}, \quad (17)$$

where k_N is the adsorption factors of nitrides, C_N themass content of nitrides, t_c the relative detention time of catalyst, $F_{c/o}$ the catalyst-to-oil ratio in the feedstock.

(c)Polycyclic aromatic adsorption:

$$F(A) = \frac{1}{1 + k_A(C_A + C_R)}, \quad (18)$$

where k_A is the adsorption factor of aromatics, C_A the mass content of aromatics, and C_R the mass content of resins in the feedstock.

(d)Arrhenius' equation:

$$k_r = k_r^0 \exp\left(\frac{E_r}{RT}\right) \quad (19)$$

(e)Arrhenius equation for any temperature, dependent on the hold-up of solids:

$$k_c\left(T,\varepsilon_s\right) = k_{r,550°C}\left(\rho,\varepsilon_s\right)\exp\left[-\frac{E_r}{R}\left(\frac{1}{T} - \frac{1}{550°C}\right)\right] \quad (20)$$

SIMULATION

The system of governing equations, twelve-lump catalytic cracking kinetic model, solid influence, and catalyst deactivation functions was solved by employing the finite volume method technique using the commercial software ANSYS CFX 14.0. The relevant results and the calculations steps are analyzed and discussed in detail in the following sections.

Geometry and Grids Generation

Steam or fuel gas is often used to lift the catalyst to the feed injection. In most designs that incorporate a "Wye" section for delivering the catalyst to the feed nozzles, a lift gas distributor is used, providing sufficient gas for delivery of dense catalyst to the feed nozzles. In other designs, the lift gas rate is several magnitudes greater, with the intent of contacting the gasoil feed into a more dilute catalyst stream. In this work the geometry of the riser is considered according to industrial reactor specifications taken from Alvarez-Castro [13] as shown in Figure 1 which reports a typical riser with Wye section.

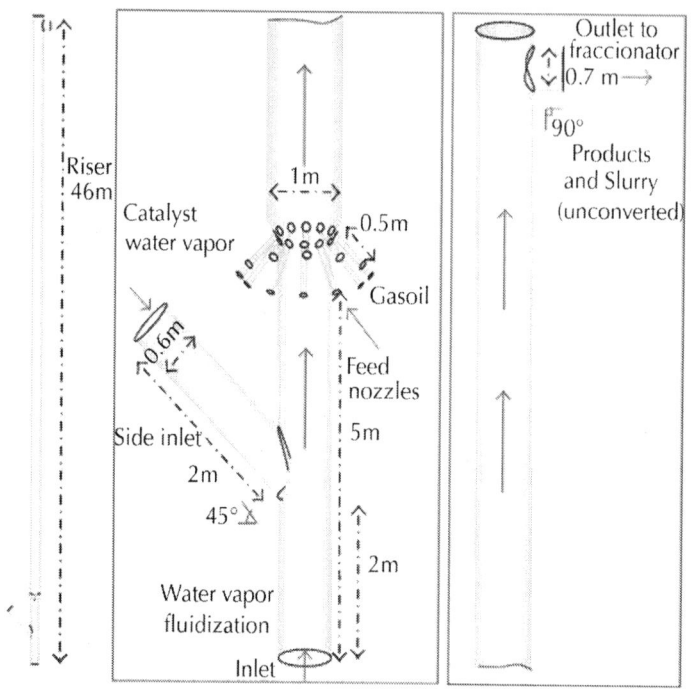

Figure 1: Riser geometry.

The geometries considered are meshed according to the procedure described above; previous works [4, 19] showed that the CFD utilized there and adopted in this work is mesh independent and meshes of 700 to 900 thousand control elements are recommended for a good representation of industrial risers. A hybrid mesh with 800 thousand control elements was built and applied in this work. Details of outlet and inlet mesh can be seen in Figure 2.

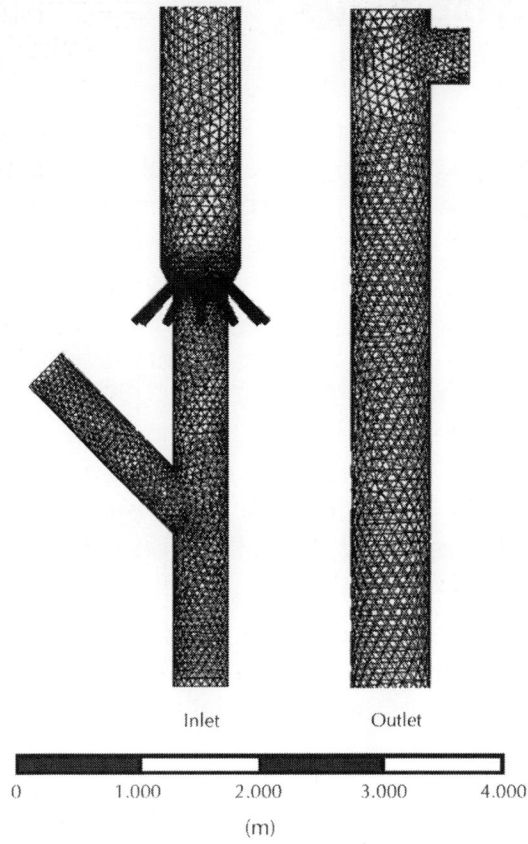

Figure 2: Mesh details.

Model Setting-Up

To implement the numerical simulation, the hydrodynamic configuration of the model was set up first and then the 12-lump kinetic model was linked with the hydrodynamic equations. Appropriate specific subroutines, that is, user defined function (UDF), were implemented in the model and solved in the CFX code in order to consider the heterogeneous, endothermic kinetics and catalyst deactivation.

Hydrodynamic Setup

The setup considered in this work considered steam as the fluidization agent which was fed into the bottom of the riser; a side inlet was used for feeding in the particle phase. A small amount of the steam (3 to 7 wt. % of the total steam) was fed together with the catalyst and 12 nozzles, 5 meters above the riser base, were used to feed gasoil; the zone where the nozzles are located is a very significant one since it is responsible for guaranteeing fast vaporization of the liquid gasoil; recent technologies have led to development of high-efficient nozzles [20–22], which implies a time for complete vaporization of about 3% (around 0.05 to 0.2 seconds) of the total reactant residence time in the reactor, in typical operation conditions. In the present simulation it was assumed that the feedstock is totally vaporized. The nonslip and free slip condition at the walls was used for the phases.

Gasoil properties and operating conditions used in the present work were taken from Wu et al. [6] and Chang et al. [14] and are summarized in Tables 1 and 2, respectively.

Table 1: Operating conditions

Item	Value
Reaction temperature (K)	793.15
Reaction time (s)	3.22
Flux of fresh feedstock (t/h)	124.46
Inlet temperature of fresh feedstock (K)	543.15
Catalyst temperature at riser inlet (K)	913.15
Ratio of catalyst to oil	8.1

Table 2: The property of feedstock

Item	Valor
Density (kg/m³ at 293.15 K)	924
Hydrogen content (wt%)	12.1
Group analysis (wt%)	
Saturates	66.05
Aromatics	25.25
Resins + asphaltenes	8.7
Distillation (K)	
HK	<578.15
10%	664.15
30%	709.15
50%	734.15
70%	771.15
Alkaline nitrides content (mg/g)	1750
Conradson carbon residue (wt%)	2.33

According to Nayak et al. [5], 400 kJ/kg is the heat to be adopted in the simulation needed for the evaporation of the liquid droplets.

Kinetic Model Setup

A 12-lump model was used to represent the products and feedstock behavior [23]. Such model can undergo a large number of reactions (56 reactions) leading to a large number of products depending on the different types of feedstock. The kinetic paths are shown in Figure 3 and Table 3 summarizes the different ranges of products and the feedstock characterization. The values of the kinetic constants, activation energies, and catalyst deactivation constant are listed in Table 4. In heat transfer model (9), Q_R is estimated by the amount of coke produced in cracking reactions; this factor Q_R is equal to 9.127103 kJ multiplied by the mass of coke which is corresponding to endothermic reactions in riser of FCC [6, 23].

Table 3: Lumps of the 12-lump kinetic model [6]

Lump symbol	Lump	Boiling range
S_S	Saturates in feedstock	613.15 K+
S_A	Aromatics in feedstock	
S_R	Resin and asphaltene in feedstock	
D_I	Diesel without pretreating LCO	477.15–613.15 K
G_S	Saturates in gasoline	C5 - 477.15 K
G_O	Olefins in gasoline	
G_A	low carbon alkanes	C3 + C4
L_n	Aromatics in gasoline	
L_{O3}	Propylene	
L_{O4}	Butene	
D_R	Dry gas	C1 + C2 + H2
C_K	Coke	

Table 4: Kinetics constants and activation energies of reaction [6, 14]

Reaction path	Activation energy (kJ/mol)	Exponential factor (m³/kg/s)	Reaction path	Activation energy (kJ/mol)	Exponential factor (m³/kg/s)
1 → 4	2.7251	0.5496	4 → 6	13.0832	0.00785
1 → 5	0.9432	0.1478	4 → 7	12.2401	0.04245
1 → 6	1.2912	0.728	4 → 8	8.2513	0.00853
1 → 7	7.2956	0.01707	4 → 9	3.7156	0.00294
1 → 8	13.0682	0.00221	4 → 10	3.4688	0.00479
1 → 9	8.9495	0.00824	4 → 11	16.3068	0.00765
1 → 10	7.787	0.00289	4 → 12	10.2884	0.06204
1 → 11	8.8766	0.02903	5 → 6	14.641	0.01914
1 → 12	9.5764	0.02268	5 → 7	15.7166	0.00595
2 → 4	4.7964	0.5068	5 → 8	15.3998	$1.06E - 05$
2 → 5	4.0451	0.09092	5 → 9	13.124	0.00982

$2 \rightarrow 6$	14.1004	0.0178	$5 \rightarrow 10$	12.8934	0.04039
$2 \rightarrow 7$	13.5735	0.02794	$5 \rightarrow 11$	18.2895	0.00547
$2 \rightarrow 8$	0.7088	0.03926	$5 \rightarrow 12$	19.805	0.00055
$2 \rightarrow 9$	3.4203	0.0579	$6 \rightarrow 5$	12.6572	0.06655
$2 \rightarrow 10$	3.7921	0.02698	$6 \rightarrow 7$	8.9658	0.1029
$2 \rightarrow 11$	4.7483	0.02206	$6 \rightarrow 8$	13.5233	$1.25E - 13$
$2 \rightarrow 12$	3.3867	0.04335	$6 \rightarrow 9$	12.1083	0.0297
$3 \rightarrow 4$	10.1081	0.04164	$6 \rightarrow 10$	12.1945	0.0246
$3 \rightarrow 5$	14.3479	0.02781	$6 \rightarrow 11$	14.6554	0.01485
$3 \rightarrow 6$	15.8237	0.1043	$6 \rightarrow 12$	11.3696	0.00878
$3 \rightarrow 7$	16.01057	0.01088	$7 \rightarrow 8$	14.0169	$1.30E - 06$
$3 \rightarrow 8$	0.9537	0.3375	$7 \rightarrow 9$	11.9348	0.01566
$3 \rightarrow 9$	1.9214	0.1208	$7 \rightarrow 10$	10.4221	0.08629
$3 \rightarrow 10$	1.35212	0.08769	$7 \rightarrow 11$	10.2512	0.09008
$3 \rightarrow 11$	4.0009	0.05663	$7 \rightarrow 12$	9.3636	0.05506
$3 \rightarrow 12$	3.9143	0.06459	$8 \rightarrow 11$	30.3051	0.002563
$4 \rightarrow 5$	14.4455	0.006942	$10 \rightarrow 11$	38.5004	0.000683
$k^A = 0.003854$		$k^N = 0.002009$		$\alpha = 0.002543$	

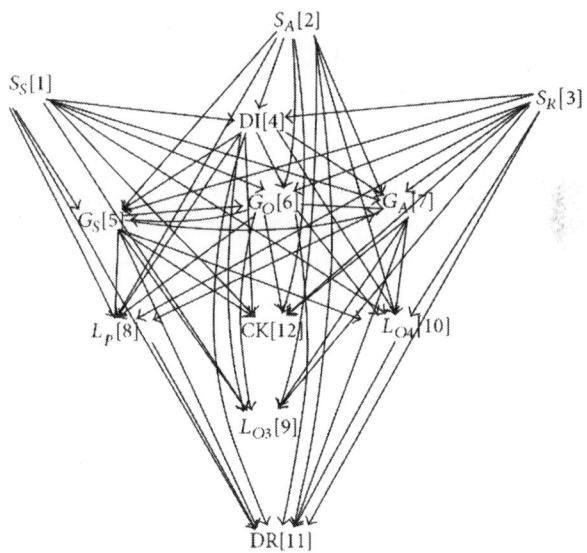

Figure 3: Twelve-lump kinetic model network [6].

Convergence

Transient expressions were estimated via the second-order backward Euler method. The convective terms were interpolated through a second-order upwind scheme "high-resolution method."

In the simulation was used a time step of 10^{-3} seconds to provide a lower Courant number in order to ensure simulation results were not dependent on the time step selected and monitoring the simulation with Courant number less than one. The convergence for progressing in time implied a residual square mean less than 10^{-4}. The simulations were solved using computers provided with Xeon 3 GHz dual core processors. About twelve days of calculation was necessary to predict a period of time (15 [s]) long enough to show that the variables had a cyclic behavior.

The following section reports the numerical results aimed at evaluating how the variation of the different operation variables affects the heat transfer, the chemical reaction, and the hydrodynamic behavior of the riser.

RESULTS AND DISCUSSION

Comparing model predictions for industrial reactors with plant data is not an easy task because the computational model requires detailed information about the feedstock as well as the design and operating conditions of the industrial setups and petroleum companies normally do not release these data on industrial risers.

Validation of the Simulation Results

The catalyst distribution profile for the riser is shown in Figure 4. Figure 4(a) shows a rendering of volume for catalyst volume fractions along an axial extension for the first six meters of the riser height where it can be seen that, just after the expansion zone and the nozzles, the feedstock has reacted and consequently is produced and the gas velocity increases due to less products density, so the catalyst moves at the high velocities imposed by the gasoil injection and its distribution becomes increasingly uniform with increasing height. Catalyst distribution is shown on eight radial contour planes in an axial direction in Figure 4(b). On the first radial planes it can be observed that, just after expansion, the solid phase (dense region) tends to agglomerate at the center of riser. This is due to the high velocity of the injected gasoil, which prevents the phenomenon of catalyst agglomeration on the walls known as coral annulus and guarantees keeping a much more uniform distribution (better homogenization) throughout the riser.

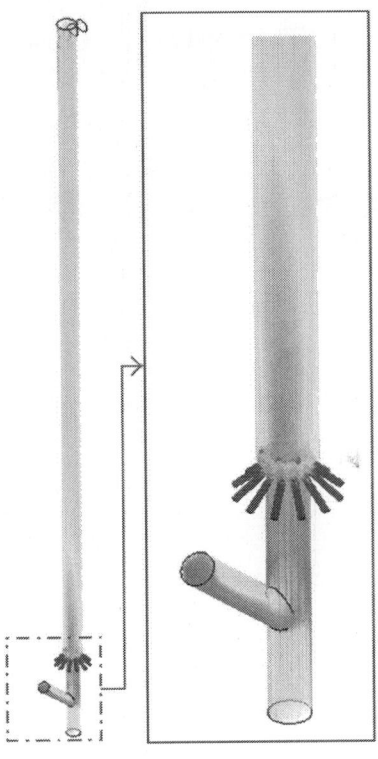

(a)

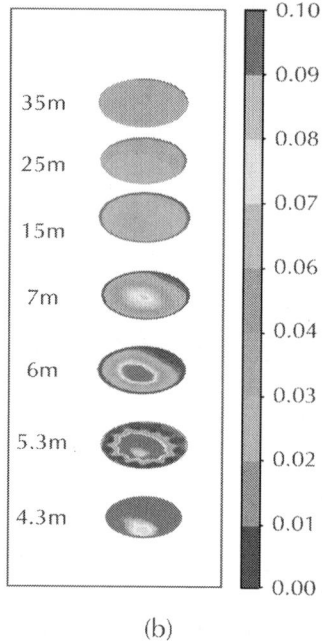

(b)

Figure 4: Volume rendering and contour profiles for axial and radial planes of catalyst volume fractions.

The fluidization velocity of the steam, at the bottom of the equipment, has a major effect on catalyst residence time in the reaction system as presented in previous work by Alvarez-Castro [13].

The model developed for the riser simulation was used to simulate the plant data reported by Chang et al. [14], in order to validate the model and compare the product yields and conversions behavior. Products distributions, that is, the average yields, along the height of the riser are shown in Figure 5. Red and green curves represent the main yield products (gasoline and diesel, resp.); it can be seen that after 25 meters an asymptotic behavior is achieved at the end of the equipment, with less conversion, due to overcracking. Blue, yellow, and brown curves represent LPG, dry gas, and coke, respectively. The black curve shows the total unconverted slurry. Results show good agreement between simulation and experimental data.

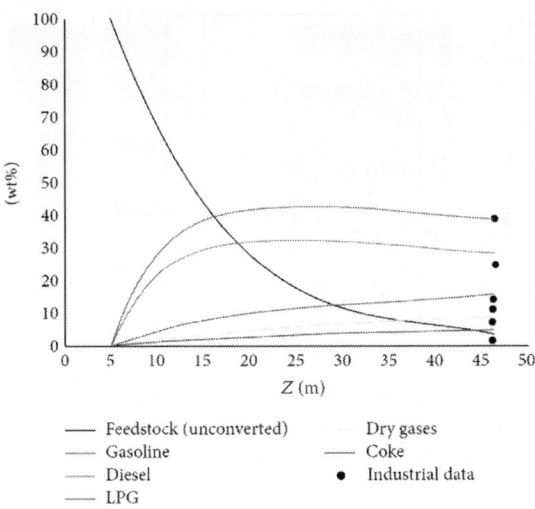

Figure 5: Model simulation results and industrial data.

Conversions and final products yields simulations model and the industrial data are shown in Figures 6 and 7, respectively. Measurements were taken at the riser outlet in order to compare the accuracy of the model simulation with the predicted results.

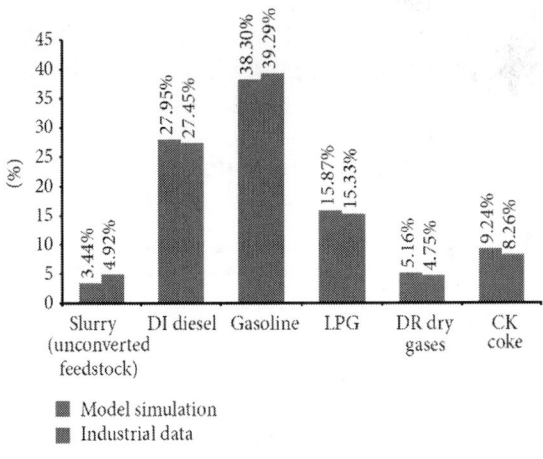

Figure 6: Comparison between product yields industrial data and the simulation model.

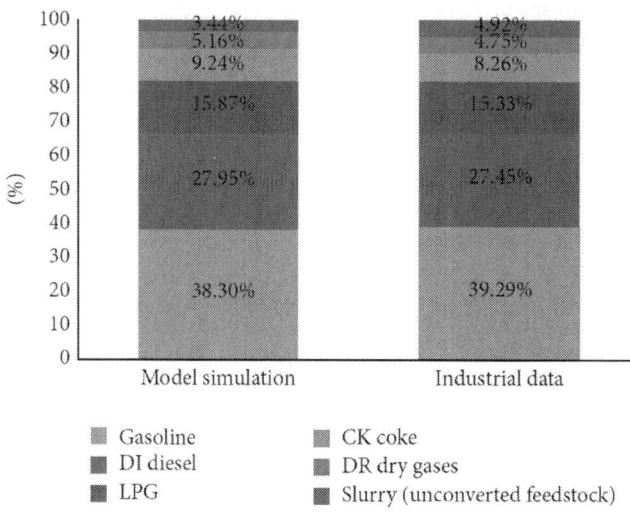

Figure 7: Comparison between industrial data and the simulation model for each case.

Operational Variables

Data obtained from Petrobras on the multipurpose pilot unit U-144 (height of 17 m and diameter of 0.52 m) in which different tests were carried out by changing the feedstock temperature, the catalyst temperature, and the catalyst-to-oil ratio are reported in Table 5.

Table 5: General behavior of the multipurpose pilot unit U-144 studied

Item	Catalyst to oil ratio	Catalyst temperature	Temperature of fresh feedstock	Residence time
	7.8 to 8.6	**680 to 720 (K)**	**530 to 550 (K)**	**1 to 2.2[s]**
Slurry (unconverted)	Decrease	Decrease	Decrease	Decrease
DI diesel	Decrease	Decrease	Decrease	Decrease
Gasoline	Decrease	Decrease	Decrease	Decrease
LPG	Increase	Increase	Increase	Increase
DR dry gas	Increase	Increase	Increase	Increase
CK coke	Increase	Increase	Increase	Increase

The sensitivity of the conversions and products yields to process variables, based on the validated simulation model, was studied. The conversions and yields were found to be very sensitive to variations in feedstock temperature, catalyst temperature, and catalyst-to-oil ratio; the differences in the conversion and product yields were in the range of 1% to 5%. Followed, results obtained for the three variables studied in this order:

- feedstock temperature
- catalyst temperature
- catalyst-to-oil ratio

Feedstock Temperature

Different case studies for temperatures ranging between 443.15 K and 643.15 K were tested while holding the other operating conditions constant, as shown in Table 6.

Table 6: Operating conditions with variations in feedstock temperature

Item	Case A	Case B	Case C	Case D	Case E
	Value	Value	Value	Value	Value
Reaction temperature (K)	793.15	793.15	793.15	793.15	793.15
Fluidization steam (%)	3	3	3	3	3
Flux of fresh feedstock (t/h)	124.46	124.46	124.46	124.46	124.46
Inlet temperature of fresh feedstock (K)	443.15	493.15	543.15	593.15	643.15
Catalyst temperature at riser inlet (K)	913.15	913.15	913.15	913.15	913.15
Ratio of catalyst to oil	8.1	8.1	8.1	8.1	8.1

(1) Comparison of the Hydrodynamics Profiles for Different Feedstock Temperatures. The global temperature (two phases) was calculated as arithmetic average contour planes for all the case studies as shown in Figure 8; the profile for case A has the lowest

inlet feedstock temperature and profile for case E has the highest. It can be observed that the temperature distributions are similar in all cases with an approximate variation of 50 [K] between the first and last cases A and E.

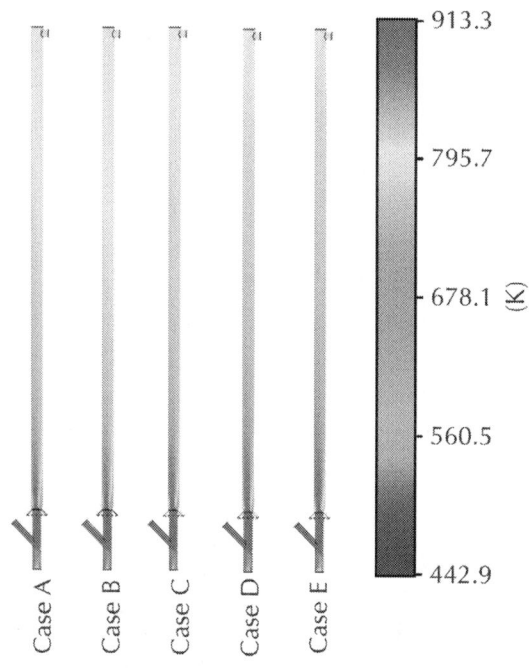

Figure 8: Temperature profiles for the axial plane with variation feedstock temperature.

Figure 9 contains the profiles for average temperature (two phases) along the center line of riser height which was also calculated as arithmetic average; the temperature decreases significantly after the feeding area, due to the endothermic nature of the reaction.

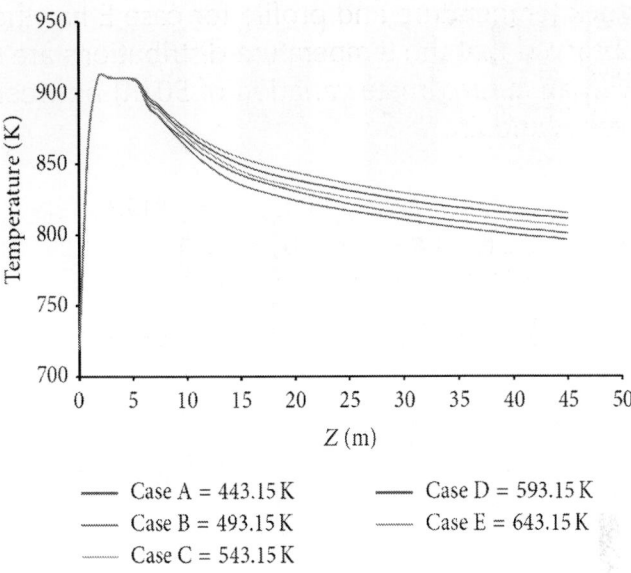

Case A = 443.15 K   Case D = 593.15 K
Case B = 493.15 K   Case E = 643.15 K
Case C = 543.15 K

Figure 9: Temperature profiles through riser with variation in feedstock temperature.

(2) Dependence of Product Yield on Feedstock Temperature. The percentage of yield and conversion products for each case presented in the previous section is shown in Figure 10. The yields were broken down into the following main groups: gasoline, diesel, LPG, dry gas, and coke. Feedstock cracking is represented by complex series-parallel reactions where gasoline and diesel are intermediate products from which the final products (LPG, dry gas, and coke) are produced. If feedstock rate of conversion is too high because of high temperature, the secondary reactions of the intermediate products cause the rate of yield to decrease due to overcracking or generation of more final products.

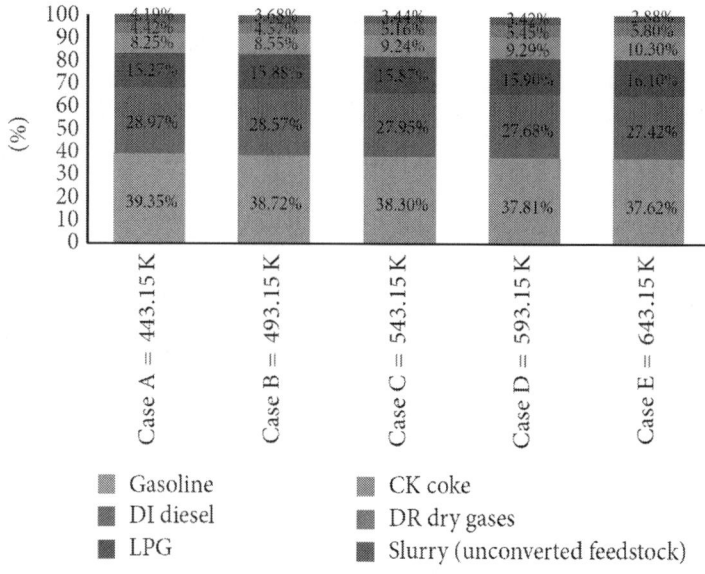

Figure 10: Products yields for each feedstock temperature.

Feedstock temperature has an important role in the process. A comparison of the product yields and conversion for all cases studied is reported in Figure 11, where it can be seen that cases A, B, and C have higher gasoline and diesel yields but a lower feedstock conversions, while cases D and E have lower gasoline and diesel yields but a higher diesel conversion. The temperature is lower at the higher gasoline and diesel yields, the importance of which should be evaluated by a cost analysis of feedstock reprocessing or production of dry gases and coke in order to improve the plant targets.

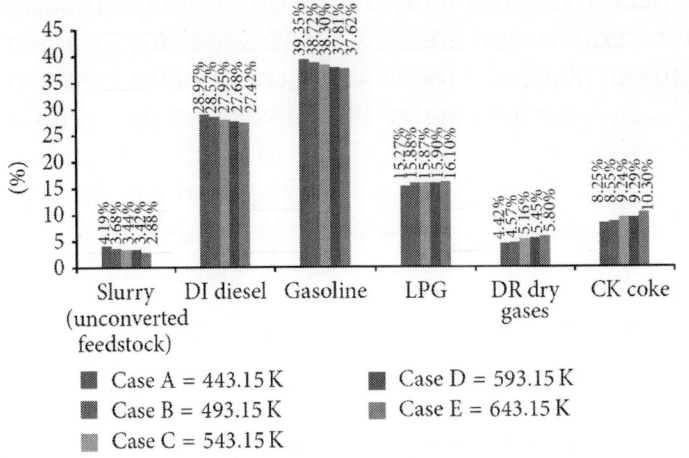

Figure 11: Comparison of the product yields for different feedstock temperatures.

Catalyst Temperature

Different cases with catalyst temperatures ranging between 813.15 [K] and 1013.15 [K] were tested while holding constant the other operating conditions as shown in Table 7.

Table 7: Operating conditions with variations in catalyst temperature

Item	Case A	Case B	Case C	Case D	Case E
	Value	Value	Value	Value	Value
Reaction temperature (K)	793.15	793.15	793.15	793.15	793.15
Fluidization steam (%)	3	3	3	3	3
Flux of fresh feedstock (t/h)	124.46	124.46	124.46	124.46	124.46
Inlet temperature of fresh feedstock (K)	543.15	543.15	543.15	543.15	543.15
Catalyst temperature at riser inlet (K)	813.15	863.15	913.15	963.15	1013.15
Ratio of catalyst to oil	8.1	8.1	8.1	8.1	8.1

(1) The Effect of Catalyst Temperature on Riser Hydrodynamics. The global temperature (gas and solid) was calculated as arithmetic average contour planes for the different case studies are shown in Figure 12. Case A is characterized by lower average overall temperature in the riser, while cases B, C, D, and E show a drastic increase in the average overall temperature in the riser with higher temperature in the profile for case E. It may be noted that small changes in the catalyst feed temperature cause a significant increase in the overall temperature.

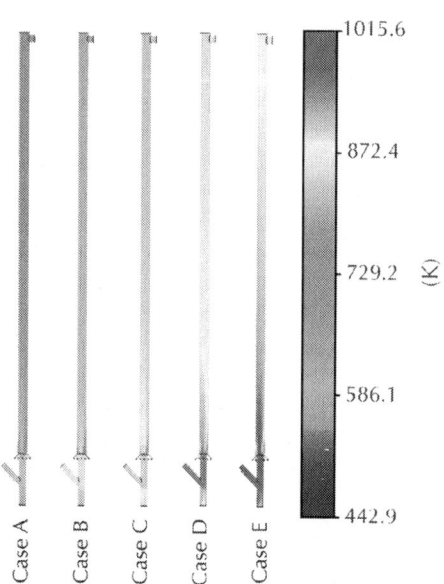

Figure 12: Global temperature profiles for the axial plane with variations in catalyst temperature.

Temperature profiles plotted along the riser height are shown in Figure 13 for all cases studied and were calculated as arithmetic average (gas and solid phases). It can be observed that catalyst temperature has a strong effect on the overall temperature in the riser, showing that the temperature profiles with a variation of 50 [K] similar to the inlet temperature of the catalyst have a much greater effect.

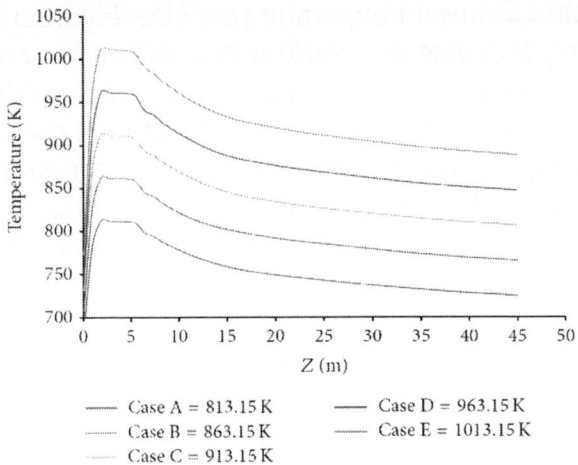

Figure 13: Temperature profiles through riser altering catalyst temperature profiles for the riser with variations in catalyst temperature.

(2) Dependence of Product Yields on Catalyst Temperature. The percentages of conversions and product yields for each case studied are shown in Figure 14. The percentages of converted gasoil and product yields are reported.

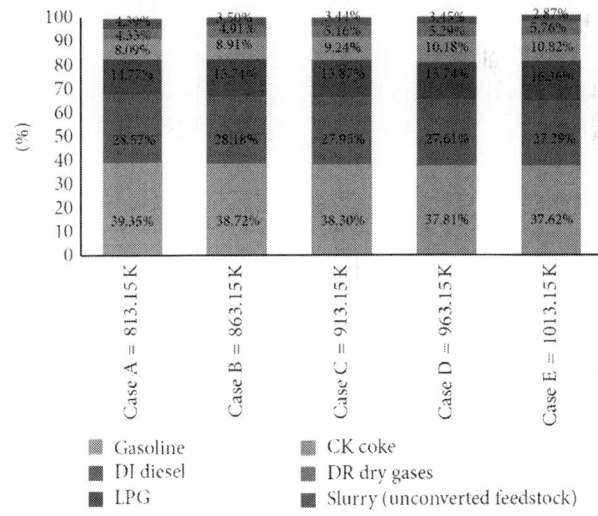

Figure 14: Products yields for each catalyst temperature.

The product yields for each case studied are shown in Figure 15. Case A has higher gasoline and diesel yields but a lower conversion of diesel, while case E has lower gasoline and diesel yields and a higher percentage of final products such as light gases, coke, and LPG. In the latter case the feedstock conversion is higher due to the higher temperature, which causes the intermediates to undergo overcracking generating lighter products of lower commercial value.

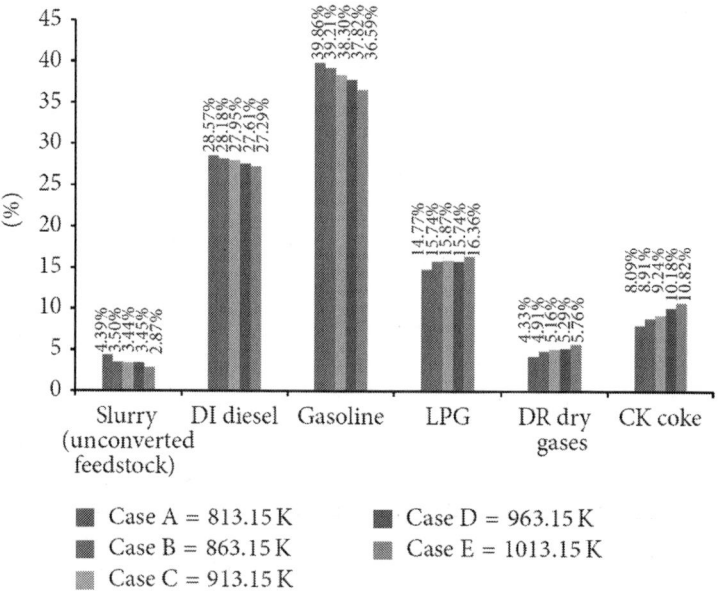

Figure 15: Comparison of the product yields for different catalyst temperatures.

Catalyst-to-Oil Ratio Study

Catalyst-to-oil ratios from 6.1 to 10.1, with step increases ratio of 1 for all cases, were studied while holding all other variables constant as shown in Table 8. Table 8: Operating conditions with variations in catalyst-to-oil ratio.

• Dependence Riser Hydrodynamics on Catalyst-to-Oil Ratio. The catalyst-to-oil ratio is an important variable, since it has a direct effect on the conversion and selectivity of gasoline and diesel. Figure 16 shows the profile of the catalyst volume fraction for the different case studies with case A having a lower catalyst-to-oil ratio and case E having a higher one in comparison to all cases studied. In both cases A and B, it can be noted that the fraction of catalyst is lower along the riser height and higher at the side where the catalyst is fed in. In cases D and E catalyst fraction is higher and more uniform along the riser height.

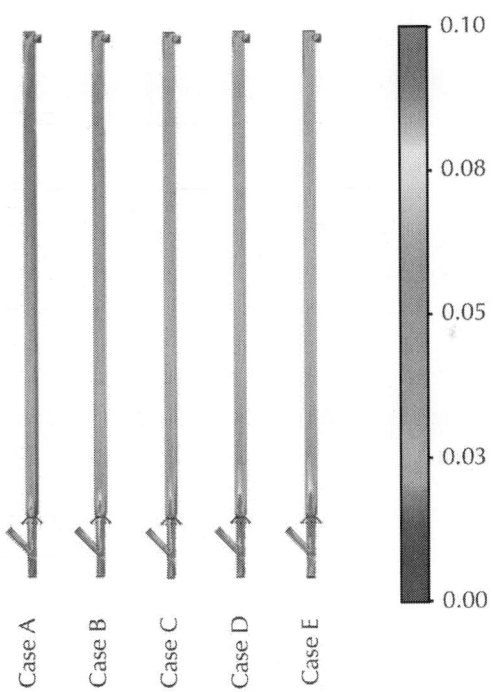

Figure 16: Catalyst volume fraction profiles for catalyst-to-oil ratio.

At the bottom of the riser, where the feedstock is injected, the temperature profile is very complex and chaotic due to the contact between hot catalyst, reagents, and steam.

Figure 17 shows the temperature profiles for a contour plane. Case A is characterized by a lower catalyst-to-oil ratio while case E represents a higher catalyst-to-oil ratio. The temperature profiles increase from case A to case E.

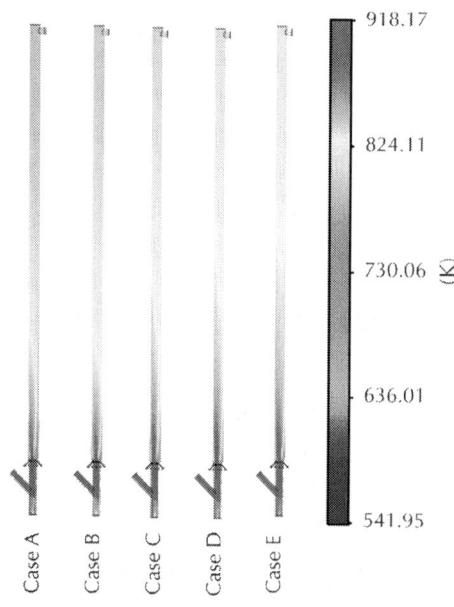

Figure 17: Temperature profiles for the axial plane with variations in catalyst-to-oil ratio.

Figure 18 contains the temperature global profiles (gas and solid) along the center line of the riser which can be observed with variations in catalyst-to-oil ratio. When the gas encounters the barrier formed by the catalyst particles, which begins the reaction, the temperature decreases slowly along the riser due to the endothermic nature of the reaction.

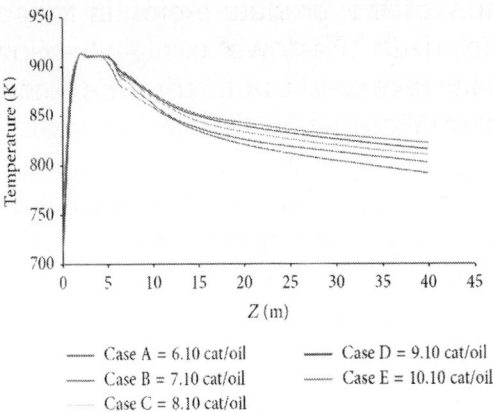

Figure 18: Temperature profiles for the riser with variation in catalyst-to-oil ratio.

- Dependence of Product Yields on Catalyst-to-Oil Ratio. The conversions and yields for each case study are reported in Figure 19 with case A characterized by a lower catalyst-to-oil ratio and case E by a higher catalyst-to-oil ratio. It can be noted that this variable has a large impact on product yield, especially for gasoline and diesel.

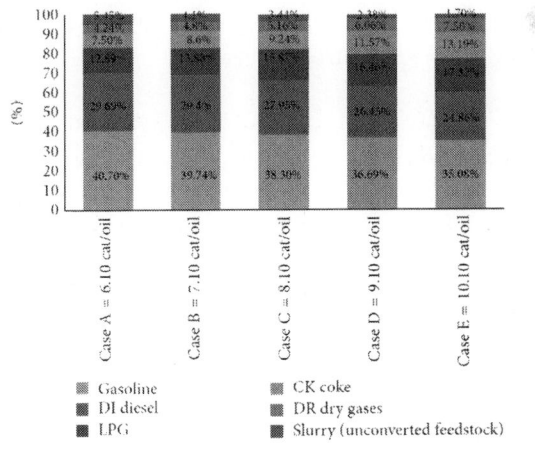

Figure 19: Product yields for each catalyst-to-oil ratio.

A comparison of the product yields in the case studied is presented in Figure 20. Case A has higher gasoline and diesel yields but a lower conversion of feedstock; on the contrary, case E, with a higher catalyst-to-oil ratio, has lower gasoline and diesel yields but higher percentages of light gases, coke, and LPG. Case A has a higher kinetic gas, but on the other hand, diesel yield is low because of a lower catalyst-to-oil ratio. Higher catalyst-to-oil ratio undergoes an overcracking, which generates lighter and lower value products.

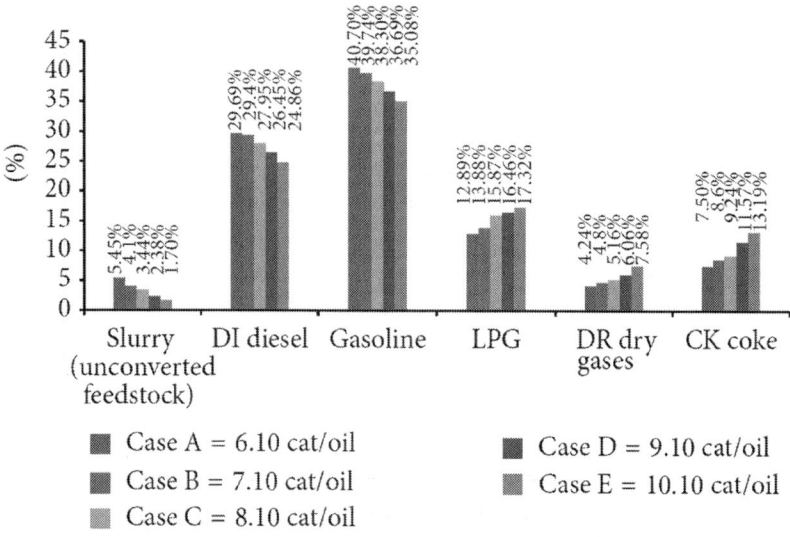

Figure 20: Comparison of the product yields for catalyst-to-oil ratio.

CONCLUSIONS

The kinetic and hydrodynamic behavior of the riser in a FCC process has been simulated employing the 12-lump kinetics model in conjunction with ANSYS CFX 14.0 software. The model has been validated against industrial data showing the ability to capture the relevant features characterizing the industrial FCC riser behavior. Systematic investigations have been carried out to study the influence of the catalyst temperature, the feedstock temperature, and the catalyst-to-oil ratio on the riser performance. Specifically, it has been shown that when the inlet conditions (of the feedstock and catalyst) are fixed, the yield of the wanted product can be increased by controlling the temperature of the riser and the catalyst-to-oil ratio. Conditions which lead to a better homogenization of the flow, avoiding unwanted hydrodynamic features, such as the core-annulus flow, which could lead to poor conversion, have also been identified. By comparing the simulated results with the experimental data, it can be concluded that the model mimics well the process; therefore, the model can be employed as a tool helping the design, operation, and control of industrial FCC risers.

ACKNOWLEDGMENTS

The authors are grateful for the financial support of Petrobras for this research.

REFERENCES

1.   R. Sadeghbeigi, "Process description," in Fluid Catalytic Cracking Handbook, R. Sadeghbeigi, Ed., chapter 1, pp. 1–42, Butterworth-Heinemann, Oxford, UK, 3rd edition, 2012.

2.   F. Durst, D. Milojevic, and B. Schönung, "Eulerian and Lagrangian predictions of particulate two-phase flows: a numerical study," Applied Mathematical Modelling, vol. 8, no. 2, pp. 101–115, 1984.

3.   X. Lan, C. Xu, G. Wang, L. Wu, and J. Gao, "CFD modeling of gas–solid flow and cracking reaction in two-stage riser FCC reactors," Chemical Engineering Science, vol. 64, no. 17, pp. 3847–3858, 2009.

4.   G. C. Lopes, L. M. Rosa, M. Mori, J. R. Nunhez, and W. P. Martignoni, "Three-dimensional modeling of fluid catalytic

cracking industrial riser flow and reactions," Computers and Chemical Engineering, vol. 35, no. 11, pp. 2159–2168, 2011.

5.   S. V. Nayak, S. L. Joshi, and V. V. Ranade, "Modeling of vaporization and cracking of liquid oil injected in a gas-solid riser," Chemical Engineering Science, vol. 60, no. 22, pp. 6049–6066, 2005.

6.   F. Y. Wu, H. Weng, and S. Luo, "Study on lumped kinetic model for FDFCC I. Establishment of model,"China Petroleum Processing and Petrochemical Technology, no. 2, pp. 45–52, 2008. ·

7.   J. Ancheyta-Juárez, F. López-Isunza, E. Aguilar-Rodríguez, and J. C. Moreno-Mayorga, "A strategy for kinetic parameter estimation in the fluid catalytic cracking process," Industrial & Engineering Chemistry Research, vol. 36, no. 12, pp. 5170–5174, 1997.

8.   H. Farag, A. Blasetti, and H. de Lasa, "Catalytic cracking with FCCT loaded with tin metal traps. Adsorption constants for gas oil, gasoline, and light gases," Industrial & Engineering Chemistry Research, vol. 33, no. 12, pp. 3131–3140, 1994.

9.   I. Pitault, D. Nevicato, M. Forissier, and J.-R. Bernard, "Kinetic model based on a molecular description for catalytic cracking of vacuum gas oil," Chemical Engineering Science, vol. 49, no. 24, pp. 4249–4262, 1994.

10.   V. W. Weekman Jr., "Model of catalytic cracking conversion in fixed, moving, and fluid-bed reactors,"Industrial & Engineering Chemistry Process Design and Development, vol. 7, no. 1, pp. 90–95, 1968.

11.   L. C. Yen, R. E. Wrench, and A. S. Ong, "Reaction kinetic correlation equation predicts fluid catalytic cracking coke yields," Oil and Gas Journal, vol. 86, no. 2, pp. 67–70, 1988.

12.   R. Sadeghbeigi, "Process and mechanical design guidelines for FCC equipment," in Fluid Catalytic Cracking Handbook, chapter 11, pp. 223–240, Butterworth-Heinemann, Oxford, UK, 3rd edition, 2012.

13. H. C. Alvarez-Castro, Analysis of process variables via CFD to evaluate the performance of a FCC riser [Ph.D. thesis], Chemical Engineering Departament, University of Campinas, 2014.

14. J. Chang, K. Zhang, F. Meng, L. Wang, and X. Wei, "Computational investigation of hydrodynamics and cracking reaction in a heavy oil riser reactor," Particuology, vol. 10, no. 2, pp. 184–195, 2012.

15. T. B. Anderson and R. Jackson, "Fluid mechanical description of fluidized beds. Equations of motion,"Industrial & Engineering Chemistry Fundamentals, vol. 6, no. 4, pp. 527–539, 1967.

16. D. Gidaspow, Multiphase Flow and Fluidization: Continuum and Kinetic Theory Descriptions, Academic Press, Boston, Mass, USA, 1994.

17. F. R. Menter, "Two-equation eddy-viscosity turbulence models for engineering applications," AIAA journal, vol. 32, no. 8, pp. 1598–1605, 1994.

18. W. E. Ranz and W. R. Marshall Jr., "Evaporation from drops, part I," Chemical Engineering Progress, vol. 48, pp. 141–146, 1952.

19. H. C. Alvarez-Castro, E. M. Matos, M. Mori, and W. P. Martignoni, "3D CFD mesh configurations and turbulence models studies and their influence on the industrial risers of fluid catalytic cracking," inProceedings of the AIChE Spring Annual Meeting, Pittsburgh, Pa, USA, 2012.

20. KBR-Technology, ATOMAX-2 Feed Nozzles, 2009.

21. J. Li, Z.-H. Luo, X.-Y. Lan, C.-M. Xu, and J.-S. Gao, "Numerical simulation of the turbulent gas-solid flow and reaction in a polydisperse FCC riser reactor," Powder Technology, vol. 237, pp. 569–580, 2013.

22. L. M. Wolschlag and K. A. Couch, "New ceramic feed distributor offers ultimate erosion protection,"Hydrocarbon Processing, pp. 1–25, 2010.

23. J. Chang, F. Meng, L. Wang, K. Zhang, H. Chen, and Y. Yang, "CFD investigation of hydrodynamics, heat transfer and

cracking reaction in a heavy oil riser with bottom airlift loop mixer," Chemical Engineering Science, vol. 78, pp. 128–143, 2012.

Chapter 2

Performance Assessment of Sintered Metal Fiber Filters in Fluid Catalytic Cracking Unit

Liang Yang[1], Zhongli Ji[1], Qiaoqi Xu[1], and Hao Li[2]

[1]College of Mechanical and Transportation Engineering, China University of Petroleum, Beijing 102249, China
[2]College of Chemical Engineering, China University of Petroleum, Beijing 102249, China

ABSTRACT

A long-term test was performed in a fluid catalytic cracking (FCC) hot gas filtration facility using sintered metal candle filters. The operating temperature and pressure were maximum 55°C and 0.28 MPa, respectively. Specific particle sampling systems were used to measure the particle size and concentration directly at high temperature. The range of inlet particle concentration is from 150 to 165 mg/Nm³. The outlet particle concentration is in the range of 0.71–2.77 mg/Nm³ in stable operation. The filtration efficiency is from 98.23% to 99.55%. The inlet volume median diameter and the outlet volume median diameter of the particle are about 1 μm and 2.2 μm, respectively. The cake thickness is calculated based on the equation of Carman-Kozeny. The effects of operating parameters including face velocity, gas cleaning pressure, pulse duration, and maximum pressure drop were investigated. The optimal operating conditions and cleaning strategies were determined. The results show that sintered metal fiber filters are suitable for industrial application due to the good performance and high efficiency observed.

INTRODUCTION

Hot gas filtration from industrial processes offers various advantages in terms of improvement of process efficiency, heat recovery, and protection of plant installation. Particularly, hot gas filtration is an essential technology for pressurized fluidized bed combustion (PFBC) and integrated gasification combined cycle (IGCC), promising coal fired generation of electricity with substantially greater thermodynamic efficiencies and reduced particulate pollutant emissions [1–4]. The filtration can protect gas turbine blades from the erosion and corrosion and improve the performance of a heat exchanger connected to a steam turbine by decreasing particles deposition.

Initially, the combined cycle power generation mentioned above has driven this development, but the focus now is shifted to the chemical and process industries. Fluid catalytic cracking (FCC) is a process for converting high molecular weight into light and getting high valuable hydrocarbons through contact with a powdered catalyst at appropriate process conditions [5–7]. Basically, the FCC process includes two sections. One section is used for the cracking to take place on contact with hot catalyst particles at approximately 520°C, and another section is designed for the regeneration of the catalyst at approximately 720°C, where the carbon deposit is reduced from 1-2 wt% to 0.05–0.2 wt% by burning in air which is fed into the regenerator at about 0.3 MPa. In a gas-catalytic reaction, the major purpose of hot gas filtration is to recover waste heat and reduce particle release. In addition, the hot gas filtration can protect the downstream equipment and meet environmental standards. In gas-solid reactions, the gas must be thoroughly cleaned in order to avoid turbine blade damage, and there is a strong thermal advantage for gas turbine power generation.

It is highly required to develop new materials and advanced operating strategies for FCC process at high temperatures. Ceramic filter is one of the most promising hot gas filtration techniques. However, due to the limitations on design and materials, long-term operation of ceramic filters is still not very successful [8]. There are some fundamental limitations due to the intrinsic material properties, which have to be improved [9]. The unreliability of the ceramic filters in demonstration trials has hindered their application [10]. Sintered metal fiber filters have been successfully used in hot gas systems for many years [11]. They have also been used for hot gas filtration in various plants due to their characteristics of fracture toughness, high thermal shock resistance, and long service life [12].

The aim of this research is to evaluate the performance of high temperature sintered metal fiber filters under a variety of operating conditions in a FCC hot gas filtration facility. The effects of operating parameters were investigated in order to determine the optimal operating conditions and cleaning strategies.

MATERIALS AND METHODS

Experimental Facility

A typical layout of a FCC unit and the location of hot gas filtration facility are illustrated in Figure 1. The fluidizing velocity is very high and close to the terminal velocity of the catalyst in both the reactor and regenerator bed. At this high velocity, the exhaust gas elutriates fine particles. The output of the reactor is fed to a battery of primary and secondary cyclones, connected in series, and positioned within the reactor vessel. These cyclones return the catalyst to the reactor bed.

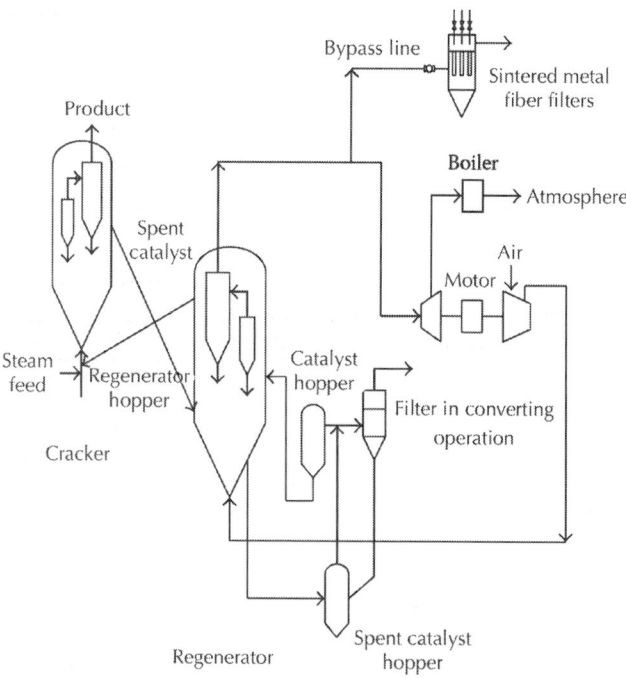

Figure 1: A typical layout of a FCC unit and the location of hot gas filtration facility.

A schematic diagram of the hot gas filtration facility is illustrated in Figure 2. The main operating parameters and composition of the FCC dusty gas are summarized in Table 1. More than 7000-hour continuous test was performed during the research. The major parts of the facility include the filtration unit with pulse cleaning, particle sampling system, and data acquisition system. The facility was designed for the maximum operating temperature of 800°C. By changing the thickness of insulation layer outside the filter vessel, the operating temperature can be controlled at the required temperature level.

Table 1: Main operating parameters and composition of the FCC dusty gas

Main operating parameters		
FCC dusty gas flow rate	$10-100\,Nm^3/h$	
Operating temperature	200–550°C	
Operating pressure	0.28 MPa	
Cleaning gas	Nitrogen	
Gas cleaning temperature	250°C	
Gas cleaning pressure	0.3–0.7 MPa	
Pulse duration	120–350 ms	
Maximum pressure drop	2–6 kPa	
Composition of the FCC dusty gas		
O_2 4.15%	CO_2 9.55%	CO 3 ppm
NO 128 ppm	NO_2 0.8 ppm	NO_x 129 ppm
SO_2 20 ppm	H_2 3 ppm	H_2S 6.4 ppm
Dew point of FCC dusty gas	86.3°C	

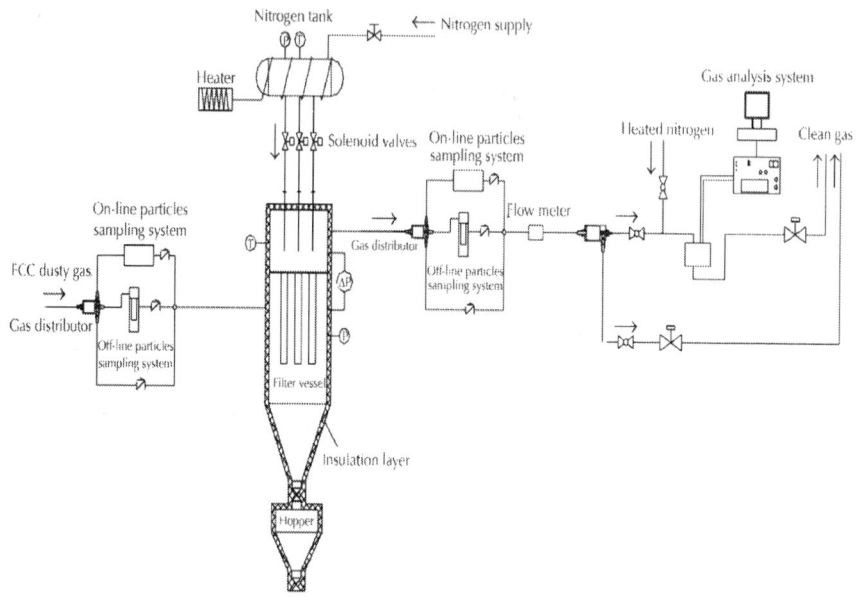

Figure 2: Schematic diagram of the hot gas filtration facility.

The filter vessel is a stainless steel cylindrical column with 360 mm diameter, 3300 mm height, and a conical base. The vessel was designed to accept three candle filters with 60 mm outer diameter and 600 mm overall length. Each filter is suspended vertically at a tube plate with 415 mm diameter and 20 mm thickness. An overall view of the filter vessel is given in Figure 3. The FCC stream from regenerator can be entered into the test facility vessel from the bypass line through a 50 mm diameter pipe with its centre line approximately 0.2 m below the tube plate. A specific gas distributor has been designed for mixing dusty gas well. A diffuse plate is placed in the front of the inlet to avoid direct impact of dusty gas on the filers. A hopper is used to collect and discharge the particles accumulated at the bottom of the filter vessel.

(a)

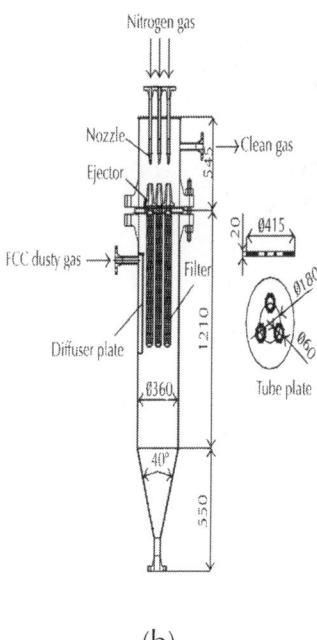

(b)

Figure 3: Overall view of the filter vessel.

Compressed nitrogen gas in the pressure range of 0.3–0.7 MPa is used for gas cleaning. The pulse cleaning duration is controlled by three solenoid valves mounted above the filter vessel. Each solenoid valve is connected to a nozzle which is directed vertically downward into a venturi injector mounted on top of the filter. The injector can also serve as a counterweight and compresses the gasket between the filter top and tube plate. The temperature of the compressed nitrogen gas was kept at about 250°C using an electronic heater to prevent condensation during the pulse cleaning. The frequency of pulse cleaning was automatically controlled by pressure in the way that the cleaning system is activated when a predetermined pressure drop is reached. The facility was connected with a computer so that the operating parameters of the system can be logged into the computer and continuously monitored during the tests.

Candle Filters

The filter elements tested during this study are sintered metal fiber filters manufactured by Bekaert Corporation. The dimensions of the filter are 60 mm outer diameter, 50 mm internal diameter, and 600 mm total length with 8 mm neck for the fixation to the tube plate. The effective filtration area is 0.113 m² for each candle filter. The porosity is about 85% and the diameters of pores are between 10 and 60 μm. The density of the filters is about 1650 kg/m³. The filters that are fabricated from AISI 430 stainless steel can resist high temperature at 1000°C.

Particle Sampling System

The particle sampling system consists of an off-line particle sampling system and an on-line particles sampling system. The two systems have been specifically designed with the aim of determining the particles concentration and particle size distribution (PSD) prior to and after the filtration vessel. The particle sampling system was specifically developed and can be used under the maximum temperature of 650°C.

The main part of the off-line particle sampling system is a sintered metal filter tube with high efficiency of 99.9% for removing 0.3 µm particles. The temperature of the particle sampling system can be controlled above 180°C by electronic heater to avoid condensation inside the sintered metal filter tube. The weight of the particles remaining in the filter tube is determined by weighing the filter tube before and after sampling. The filtration efficiencies reported below are based on the inlet gas concentrations, gas flow rate duration of sampling, and the changes in the weight of the filter tube. The on-line particle sampling system uses an intense source of white light to illuminate a nearly cubic particle sensing volume of located at the centre of the aerosol flow path. This volume is defined by a combination of apertures placed in the optical path of the illumination and the two sensing branches, which are arranged at an observation angle of 90° opposing each other [13]. The principle and more details of the on-line particles sampling system can be found in our previous research [14].

RESULTS AND DISCUSSION

Initial and Residual Pressure Drop

The initial pressure drop is defined as variation of the pressure drop across filters with the face velocity by considering clean gas through them. The influence of the temperature and face velocity on the pressure drop across the filters can be investigated during these tests. Figure 4 shows the evolution of the initial pressure drop with the face velocity at five temperature levels: 305°C, 365°C, 425°C, 495°C, and 550°C, respectively. It can be observed that the pressure drop increases approximately linearly with the face velocity, as Darcy's Law indicates. When the operating temperature increases from 305°C to 550°C, the pressure drop increases for the

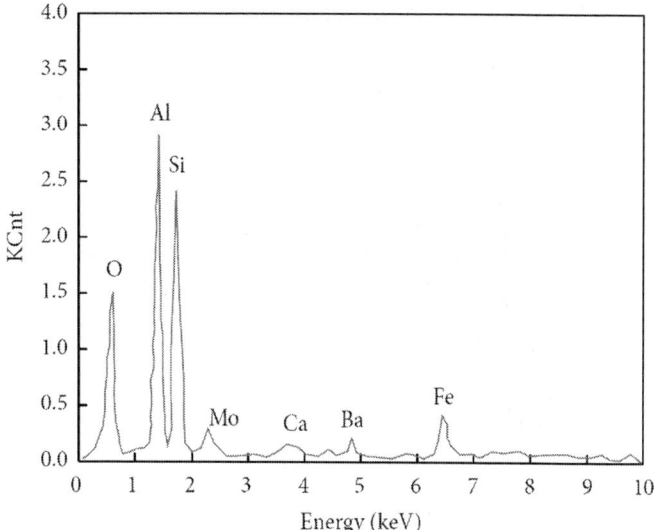

Figure 8: The image of EDX measured results.

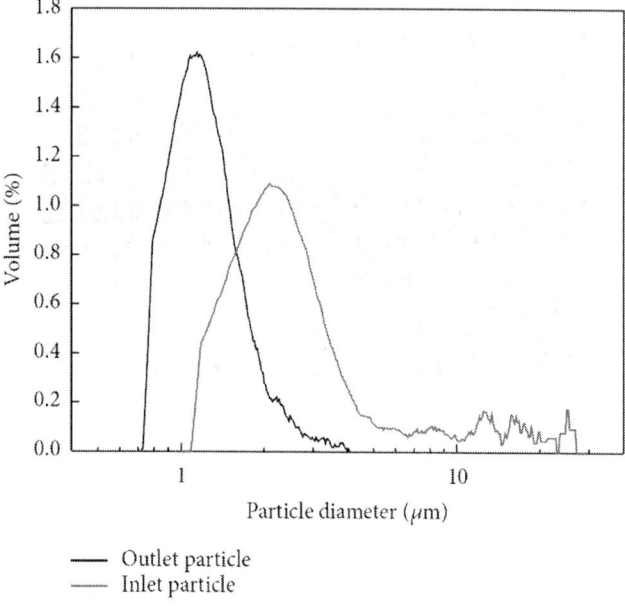

Figure 9: PSD measured by the Coulter counter analyzer.

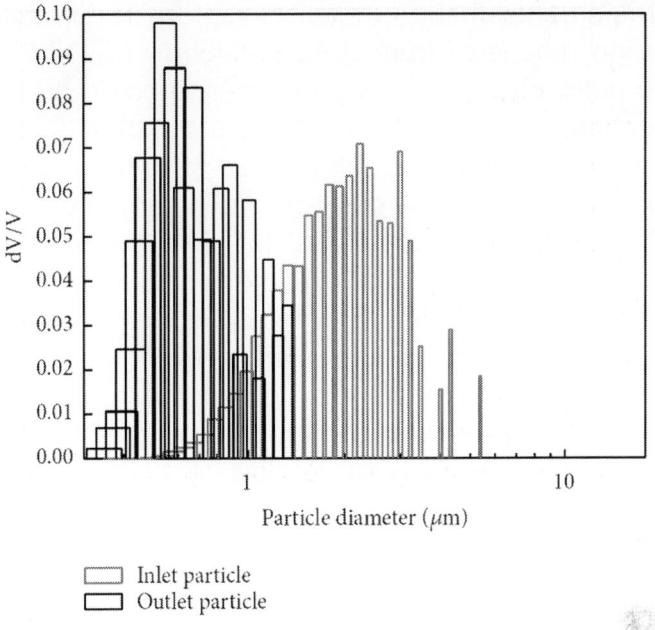

Inlet particle
Outlet particle

Figure 10: PSD measured by the on-line particle sampling system.

It can be observed from Figure 7 that the inlet particles show a good spherical shape. The inlet volume median diameter and the outlet volume median diameter of the particles measured by the off-line particle sampling system are about 1.14 µm and 2.37 µm, respectively. The inlet volume median diameter and the outlet volume median diameter of the particles measured by the on-line particle sampling system are about 0.92 µm and 2.21 µm, respectively. The results measured by off-line particle sampling system agree well with those measured by on-line particle sampling system.

Filtration Efficiency

A peak emission is found after each pulse cleaning during an unstable operation, and a typical emission profile is illustrated in Figure 11. However, the emission phenomenon disappeared after the 100th cycle. During pulse cleaning, it is found that the outlet particle concentration increases rapidly. The outlet particle concentration changes from 16.25 mg/Nm³ to 37.55 mg/Nm³ during the pulse cleaning. This phenomenon could be caused by several mechanisms such as the decrease of filtration efficiency due to the dust cake detachment, the particle penetration through the filter media due to the pulse cleaning shock, or the direct particle penetration through the filters [15–17]. One of the reasons for this problem may be the serious gas reflux from outside to inside of filters during pulse cleaning. The reflux causes fine particles to redeposit on outer surface of the filters or even penetrate into the filters. However, due to the formation of a constant and dense residual dust cake layer on the filters surface, the emission peaks progressively decrease. This residual dust cake layer dominates the subsequent filtration and stable operation.

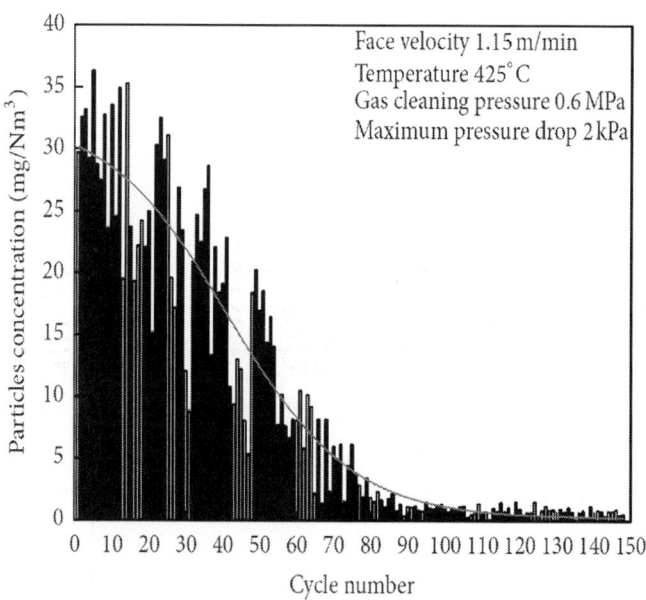

Figure 11: Typical particles emission profile during operation.

When the filters are in stable operation, the filtration efficiencies can be measured at three face velocities (1.15 m/min, 1.5 m/min, and 2.2 m/min) and three temperatures (305°C, 425°C, and 550°C).

The maximum pressure drop is 2 kPa and the gas cleaning pressure is 0.6 MPa. The test results of particle concentration and filtration efficiency are listed in Table 3. A significant effect of temperature on the filtration efficiency cannot be observed during these tests. The inlet particle average concentration range is from 150 to 165 mg/Nm³. The outlet particle average concentration is in the range of 0.71–2.77 mg/Nm³. Slower pressure drop building up during the filtration, which results in longer pulse cleaning intervals, was one of the reasons for the higher filtration efficiencies. The filtration efficiency decreases from 99.55% to 98.23% when the face velocity is increased from 1.15 m/min to 2.2 m/min. The decrease in the filtration efficiency is associated with an increase in the pulse cleaning frequency, which could eventually result in a thinner dust cake and poorer filtration.

Table 3: Test results of particles concentration and filtration efficiency

Face velocity/m/min	Temperature/°C	Inlet particles concentration/mg/Nm³	Outlet particles concentration/mg/Nm³	Filtration efficiency/%
1.15	305	162.65	1.24	99.23
	425	155.76	1.02	99.34
	550	159.43	0.71	99.55
1.5	305	158.32	1.23	99.22
	425	152.45	1.13	99.26
	550	161.23	1.37	99.15
2.2	305	156.54	2.77	98.23
	425	163.42	2.35	98.56
	550	153.25	2.02	98.68

Dust Cake Thickness

Since it is difficult to measure the cake thickness directly, the cake thickness is estimated from the measured increase of pressure drop across the dust cake using the well-known equation of Carman-Kozeny [18], which is applied to the laminar flow conditions (Re < 1), giving

$$h = \frac{\left(\Delta P_{cake}\varphi^2 d_v^2 \varepsilon^3\right)}{\left[C\mu(1-\varepsilon)^2 U\right]},$$

(1)

where h is the dust cake thickness, ΔP_{cake} is the pressure drop across the dust cake, φ is the particle sphericity, d_v is the diameter of an ideal sphere having the same volume as the particles in the experiment, ε is the cake porosity, C is the Carman-Kozeny constant (C=180), μ is dynamic viscosity of gas at filter's operating temperature, and U is face velocity. Particle diameter of 2.4 μm and particle sphericity of 0.87 (observed from the SEM image of Figure 7) are used for dust cake. The cake porosity was estimated from experimental investigations by Schmidt on dust cakes deposited on filter media [19]. Schmidt reported that, for limiting pressure drops of 0.5, 1, and 2 kPa, the dust cake porosity ranged from 85% at the cake surface to 68% at the filter surface. In this research, a cake porosity of 80% was used to estimate the cake thickness. Depending on the maximum pressure drop and the face velocity, the estimated cake thickness is in the range of 0.43–4.15 mm. The estimated cake thickness can be used to determine the cake mass as well as the percentage of dust attached to the filters. The cake mass can be calculated from the equation below:

$$m_{cake} = \rho_B L \left[\pi(R_i + h)^2 - \pi R_i^2\right],$$

(2)

Where m_{cake} is the cake mass, ρ_B is the bulk density of the particles, L is the effective candle filter length, and R_i is the outer radius of the clean filter.

The operating temperature is 425°C, the gas cleaning pressure is 0.6 MPa, and the inlet particles concentration is approximately 160 mg/Nm³. The bulk density of the particles is approximately 730 kg/m³. The results indicate that the percentage of particles attached on filters was in the range of 65–90%. At a face velocity of 2.2 m/min, nearly 86% particles can be attached to the filters. Dust cake thicknesses estimated and pressure drop across the cake are summarized in Table 4.

Table 4: Dust cake thicknesses estimated and pressure drop across the cake

Maximum pressure drop/kPa	Face velocity/m/min	Pressure drop across cake/kPa	Estimated cake thickness/mm	Estimated cake mass/g	Dust attached to filters/%
3.5	1.15	2.06	1.75	68.74	65
	1.5	1.56	1.32	54.23	72
	2.2	0.37	0.43	28.69	78
4.2	1.15	2.79	3.04	97.65	68
	1.5	2.28	2.37	85.22	78
	2.2	1.12	1.21	60.37	81
5.0	1.15	3.52	4.15	129.05	71
	1.5	3.04	3.32	110.24	82
	2.2	1.83	1.43	78.56	86

Effect of the Operational Parameters

Face Velocity

The pressure drop across the filters was primarily affected by the face velocity. Figures 12 and 13 show the pressure drop variations as a function of time for face velocities of 1.15, 1.5, 2.2, and 2.55 m/min. The maximum pressure drop was set at 5 kPa during these tests. At a face velocity of 1.15 m/min, it took about 28 h for the pressure drop to build from the initial to the maximum pressure drop. The reasons for the extremely slow increase in the pressure drop include the high bulk density of the catalyst particles, the low inlet particle concentration, and the relatively smooth and slippery surfaces of the filters, which lead to a weak adhesion of the dust cake to the filters. Similar behaviour can be observed for the face velocity of 1.5 m/min and 2.2 m/min, and it can be observed that operation is stable when filtration velocity is from 1.15 m/min to 2.2 m/min. However, unstable operation occurs at the face velocity of 2.55 m/min. During the unstable operation illustrated in Figure

13, a rapid increase in the pressure drop across the filters can be found and the slope of initial pressure drop increases continuously, leading to an unstable operation and more frequent pulse cleaning and determining a maximum value for the face velocity that should not be overcome.

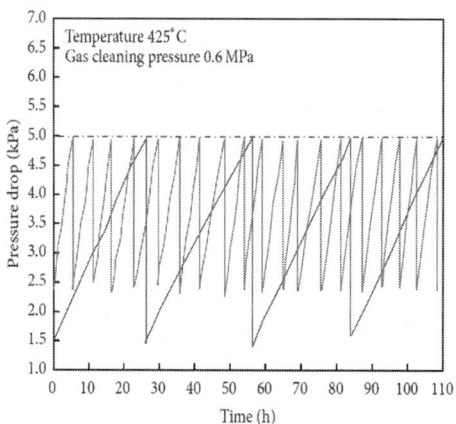

Figure 12: Pressure drop variations for face velocities of 1.15 m/min and 1.5 m/min.

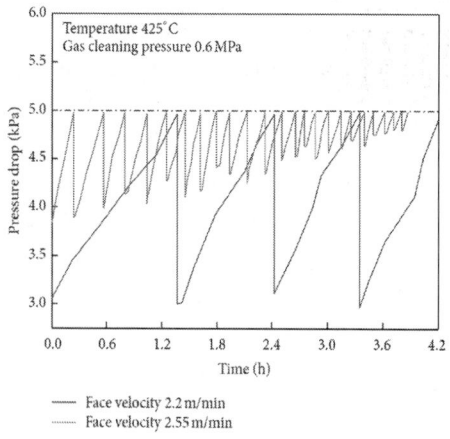

Figure 13: Pressure drop variations for face velocities of 2.2 m/min and 2.55 m/min.

Gas Cleaning Pressure

Four different gas cleaning pressures are tested under the same operating conditions. Figures 14 and 15 present the effect of the gas cleaning pressure on the pressure drop evolution. However, it must be pointed out that an upper limit of pressure cleaning can be determined, from which the frequency of the pulse remains constant despite an increase in the cleaning pressure. The improvement of pressure drop, using a cleaning pressure higher than 0.6 MPa, was not very significant; therefore, cleaning pressure value of 0.6 MPa was selected for the tests so as to minimize the consumption of nitrogen. A lower limit of gas cleaning pressure was also determined, below which the operation was not feasible. It can be observed that the operation is not stable when gas cleaning pressure is lower than 0.45 MPa. The efficient gas cleaning pressure value depends on the operating conditions but it is normally about twice the pressure value inside the filter vessel.

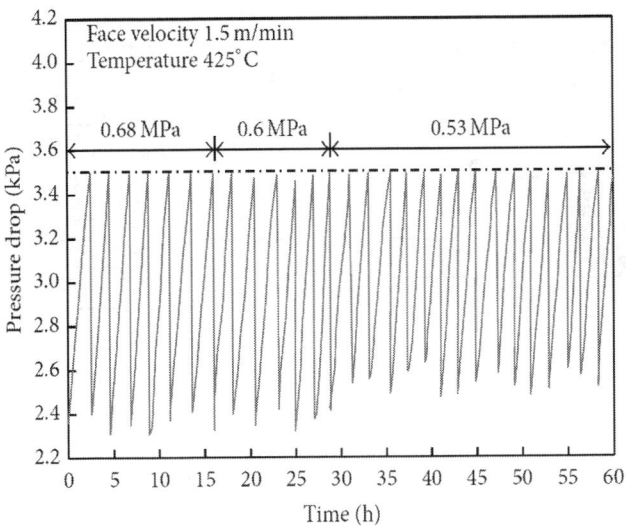

Figure 14: Pressure drop variations for gas cleaning pressure of 0.68 MPa, 0.6 MPa, and 0.53 MPa.

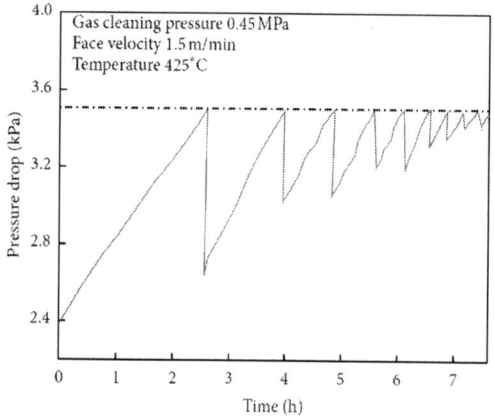

Figure 15: Pressure drop variations for gas cleaning pressure of 0.45 MPa.

Pulse Duration

Figure 16 shows the results of pressure drop variations measured at different duration times of the cleaning pulse, including 180 ms, 240 ms, and 300 ms. It can be found that the long pulse duration values of 240 ms and 300 ms do not show a significant effect on the pressure drop compared with the short pulse duration of 180 ms.

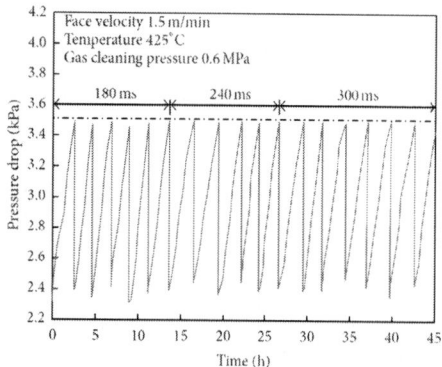

Figure 16: Pressure drop variations for pulse durations of 180 ms, 240 ms, and 300 ms.

Maximum Pressure Drop

Figure 17 shows the results of pressure drop variations as a function of time for three different maximum pressure drops at a face velocity of 1.5 m/min. In Figure 17, it can be seen that stable operation is possible at three values of maximum pressure drop. For maximum pressure drop values of 3.5 kPa and 4.2 kPa, both initial pressure drop remained constant and the same to each other. But for the maximum pressure drop of 2.8 kPa, the initial pressure drop is higher than those of the other two values. The reason for this phenomenon is that the lower maximum pressure drop results in a thin dust cake thickness which decreases the pulse cleaning efficiency. The frequency of the pulse cleaning decreases with increasing the level of maximum pressure drop. The pulse interval increased from 3 h to 5 h when the maximum pressure drop was increased from 3.5 kPa to 4.2 kPa. Thus, a 40% reduction of pulse cleaning interval can be achieved by the higher value of maximum pressure drop. However, serious deterioration of the filters may be expected since the operation becomes more severe at higher levels of maximum pressure drop.

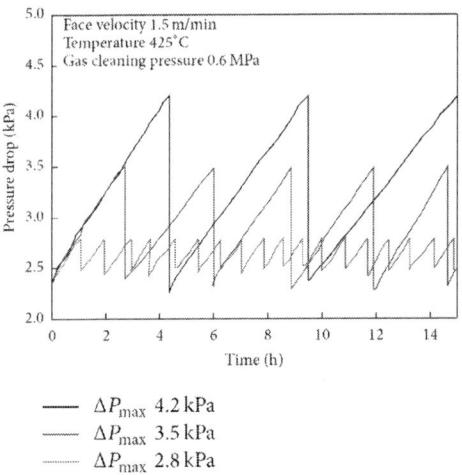

Figure 17: Pressure drop variations for maximum pressure drop of 2.8 kPa, 3.5 kPa, and 4.2 kPa.

CONCLUSIONS

A long-term test was performed in a fluid catalytic cracking (FCC) hot gas filtration facility using three sintered metal candle filters under a variety of operating conditions. The effects of operating parameters including face velocity, gas cleaning pressure, pulse duration, and maximum pressure drop were investigated.

Due to the rise of the gas viscosity with temperature, the pressure drop increases when the operating temperature increases at the same face velocity values. There are three reasons for the slow increase in the pressure drop, including the high bulk density of the catalyst particles, the low inlet particle concentration, and the relatively smooth and slippery surface of the filters, which leads to weak adhesion of the dust cake to the filters. The range of inlet particle concentration is from 150 to 165 mg/Nm^3. The outlet particle concentration is in the range of 0.71–2.77 mg/Nm^3 during the stable operation. The inlet volume median diameters and the outlet volume median diameters of the particles are about 1 μm and 2.2 μm, respectively. The filtration efficiency is more than 99% when the face velocity is kept in the range of 1.15–1.5 m/min. At a face velocity of 2.2 m/min, however, the filtration efficiency is reduced to a value lower than 99%. The cake thickness can be calculated based on the equation of Carman-Kozeny.

The pressure drop across the filters is primarily affected by the face velocity. The appropriate face velocity should not exceed 2.2 m/min in this research. Excessive filtration velocity leads to an unfeasible operation. Using gas cleaning pressure higher than 0.6 MPa was not very significant, and the operation was not feasible when the gas cleaning pressure was lower than 0.45 MPa. Pulse duration does not show a significant effect on pulse cleaning performance. The frequency of the pulse cleaning decreases with the higher level of maximum pressure drop. However, serious deterioration of the filters may be caused since the operation becomes more severe under a higher maximum pressure drop. In general, the study in this paper shows that the sintered metal fiber filter is suitable for industrial application due to the good performance and high efficiency observed in the experiments. The results are very useful for the next research. The main purposes of the next research are to reduce the emissions to the atmosphere according to the progressively more stringent emission standards and to protect downstream equipment from erosion. We will discuss the long-term operation data in detail later in the next paper.

ACKNOWLEDGMENTS

The authors acknowledge the sintered metal candle filters provided by Bekaert Corporation Belgium and the experiment assistance by PetroChina Changqing Petrochemical Company.

REFERENCES

1.   S. Ito, T. Tanaka, and S. Kawamura, "Changes in pressure loss and face velocity of ceramic candle filters caused by reverse cleaning in hot coal gas filtration," Powder Technology, vol. 100, no. 1, pp. 32–40, 1998.

2.   R. A. Newby, T. E. Lippert, M. A. Alvin, G. J. Burck, and Z. N. Sanjana, "Status of Westinghouse hot gas filters for coal and biomass power systems," Journal of Engineering for Gas Turbines and Power, vol. 121, no. 3, pp. 401–408, 1999.

3.   D. H. Smith and G. Ahmadi, "Problems and progress in hot-gas filtration for pressurized fluidized bed combustion (PFBC) and integrated gasification combined cycle (IGCC)," Aerosol Science and Technology, vol. 29, no. 3, pp. 163–169, 1998.

4.   S. Hajek and W. Peukert, "Experimental investigations with ceramic high-temperature filter media,"Filtration and Separation, vol. 33, no. 1, pp. 29–37, 1996.

5.   H. Sasatsu, N. Misawa, M. Shimizu, and R. Abe, "Predicting the pressure drop across hot gas filter (CTF) installed in a commercial size PFBC system," Powder Technology, vol. 118, no. 1-2, pp. 58–67, 2001. · ·

6.   J. P. K. Seville and R. Clift, Gas Cleaning in Demanding Applications, Edited by J. P. K. Seville, Blackie/Kluwer, London, UK, 1997.

7.   R. A. Meyers, Handbook of Petroleum Refining Processes, McGraw Hill, 2nd edition, 1996.

8.   V. Vasudevan, B. S. J. Kang, and E. K. Jonson, "A study on ash particle distribution characteristics of candle filter surface regeneration at room temperature," US DOE grant No. DE-FC26-99FT40203, 2002.

9.   S. D. Sharma, M. Dolan, D. Park et al., "A critical review of syngas cleaning technologies—fundamental limitations and practical problems," Powder Technology, vol. 180, no. 1-2, pp. 115–121, 2008.

10.   P. Kilgallon, N. J. Simms, J. E. Oakey, and I. Boxall, "Metallic filters for hot gas cleaning," Tech. Rep. Coal R239, Power Generation Technology Center, Cranfield University, Cranfield, UK, 2004.

11.   D. B. Purchas and K. Sutherland, Handbook of Filter Media, Elsevier Advanced Technology, New York, NY, USA, 2nd edition, 2002.

12.   S. Jha, R. S. Sekellick, and K. L. Rubow, "Sintered metal hot gas filters," in High Temperature Gas Cleaning, A. Dittler, G. Hemmer, and G. Kasper, Eds., vol. 2, p. 492, 1999.

13.   M. Heim, B. J. Mullins, H. Umhauer, and G. Kasper, "Performance evaluation of three optical particle counters with an efficient "multimodal' calibration method," Journal of Aerosol Science, vol. 39, no. 12, pp. 1019–1031, 2008.

14.   Q. Xu, Z. Ji, and L. Yang, "Performance assessment of high temperature flue gas on-line particle size analyzer," CIESC Journal, vol. 63, no. 11, pp. 3506–3512, 2012.

15.   E. Schmidt and F. Löffler, "Preparation of dust cakes for microscopic examination," Powder Technology, vol. 60, no. 2, pp. 173–177, 1990.

16.   X. Simon, D. Bémer, S. Calle, D. Thomas, and R. Régnier, "Description of the particle puff emitted downstream of

different dust separators consecutive to pulse-jet cleaning," Filtration, vol. 5, no. 1, pp. 52–61, 2005.

17. J. Binnig, J. Meyer, and G. Kasper, "Origin and mechanisms of dust emission from pulse-jet cleaned filter media," Powder Technology, vol. 189, no. 1, pp. 108–114, 2009.

18. C. Tien and B. V. Ramarao, "Can filter cake porosity be estimated based on the Kozeny-Carman equation?" Powder Technology, vol. 237, pp. 233–240, 2013.

19. E. Schmidt, "Experimental investigations into the compression of dust cakes deposited on filter media,"Filtration and Separation, vol. 32, no. 8, pp. 789–793, 1995.

Chapter 3

CFD Study of Industrial FCC Risers: The Effect of Outlet Configurations on Hydrodynamics and Reactions

Gabriela C. Lopes[1], Leonardo M. Rosa[1], Milton Mori[1], José R. Nunhez[1], and Waldir P. Martignoni[2]

[1]School of Chemical Engineering, University of Campinas, 500 Albert Einstein Avenue, 13083-970 Campinas, SP, Brazil

[2]PETROBRAS, 65 República do Chile Avenue, 20031-912 Rio de Janeiro, RJ, Brazil

ABSTRACT

Fluid catalytic cracking (FCC) riser reactors have complex hydrodynamics, which depend not only on operating conditions, feedstock quality, and catalyst particles characteristics, but also on the geometric configurations of the reactor. This paper presents a numerical study of the influence of different riser outlet designs on the dynamic of the flow and reactor efficiency. A three-dimensional, three-phase flow model and a four-lump kinetic scheme were used to predict the performance of the reactor. The phenomenon of vaporization of the liquid oil droplets was also analyzed. Results showed that small changes in the outlet configuration had a significant effect on the flow patterns and consequently, on the reaction yields.

INTRODUCTION

Although commercially established for over half a century, the fluid catalytic cracking (FCC) process is still widely studied nowadays. Since it is a very profitable operation, any improvement in it can result in large savings for the refinery. In the FCC process, preheated high-boiling liquid oil is injected into the riser reactor, where it is vaporized and cracked into smaller molecules by contact and mixing with the very hot catalyst particles coming from the regenerator. These phenomena cause a gas expansion, which drags the catalyst to the top of the reactor. Since catalytic cracking reactions can only occur after the vaporization of liquid feedstock, mixing of hydrocarbon droplets with catalyst must take place in the riser as soon as possible.

It is known that riser reactors have complex hydrodynamics. They present a high solids concentration near the walls and are also axially divided into dense and dilute regions. In addition, different riser configurations such as the inlet and outlet structures can have a profound effect on the flow patterns mentioned above.

The influence of riser exit geometry on the hydrodynamics of gas-solid circulating fluidized beds (CFB) has been investigated in many studies [1–6]. Although the results reported in these studies apparently conflict quantitatively concerning the influence of riser exit, some common aspects can be observed: (1) the design of the exit has a large effect upon the reflux of solids; (2) abrupt exits cause an increase in the solids holdup and a large backmixing at the top of the riser; (3) increasing the refluxing effect of the exit has proved to increase the mean particle residence time; (4) larger and denser clusters are formed at the walls in the risers with abrupt exits.

In an experimental work, Lim et al. [7] investigated a cold model of a circulating fluidized bed, in a riser with an horizontal and flat cover at the top. They limited the operating conditions, that resulted in stable operation of the circulating fluidized bed, and proposed a model to estimate the ratio of solids that exit the riser to solids that recirculate back into the riser. This model predicts the cases in which solids inertia dominates, and the cases when solids have insufficient inertia to resist the change in airflow.

The presence of reverse core-annulus profile under certain conditions in gas-solid CFBs was also observed in an experimental study by Chew et al. [8]. Although some previous works explain this behavior as a consequence of the impact of gas-phase turbulence associated with dilute flows, Chew et al. [8] suggest a dominant factor for reverse core-annulus flow: the particle Stokes number (St). According to their work, particles with large St are more likely to follow more diffuse trajectories after collision rather than following fluid streamlines, because of greater particle inertia relative to fluid viscous forces.

Van Der Meer et al. [3] defined a parameter to quantify the reflux of solids in a square cross-section riser of a laboratory CFB. They concluded that the values of this parameter obtained for the different outlet configurations vary in a factor of 25, showing that the exit design has a significant effect on riser flow regime.

Pugsley et al. [9] performed experiments with sand and FCC catalyst in two cold CFBs of 0.1 m and 0.2 m diameter in order to observe the relative influence of smooth and abrupt exit configurations on the axial pressure drop profile. They concluded that the reflux along the riser length induced by abrupt riser exits is related to the riser diameter and the particle terminal velocity. For heavier and largest particles, the influence of abrupt exits was observed to affect a longer section of the riser when compared to the effect of smaller particles. They also observed that these effects are less pronounced for smaller riser diameters.

In a recent study, Van Engelandt et al. [10] analyzed experimentally and computationally the riser outlet effects induced by an L-outlet and by abrupt T-outlets with different extension heights, outlet surface areas, and gas flow rates. They also observed that the T-outlet configuration is found to induce recirculation by vortex formation in the extension part of the riser and cause a main reflux at the wall opposite to the riser outlet. Moreover, a reduction of the outlet surface area of a T-outlet results in an increased solids holdup in the extension part of the riser. With the increase of the gas flow rate, the position of the vortex and the anisotropy of the fluctuating particle velocities are strongly affected.

A numerical study about the effects of different outlet surface area on the flow pattern was also performed by De Wilde et al. [11]. The results obtained in their simulations showed the need of performing 3D calculations in order to predict accurately the exit effects of abrupt outlet configurations. Chalermsinsuwan et al. [12] studied different designs of the riser geometries based on the improvement of main factors that have an effect on combustion, gasification, and cracking reaction characteristics, using a 2D transient Eulerian approach.

Das et al. [13] used CFD simulations to investigate the effect of different flow and design parameters on the adsorption reactions in a SO_2–NO_x riser. They showed that the solids recirculation at the top section of the riser, induced by abrupt T outlets significantly, decreases the NO and NO_2 removal, worsening the reactor efficiency. They emphasized that this analysis is just possible due to the use of 3D models, which allows predicting the effects of outlet geometries on the flow and reaction fields.

These works indicate that the hydrodynamics of CFB risers is extremely complex, showing that the distribution of axial solids concentration depends not only on operating conditions such as gas velocity, solids flux, and particle properties, but also on the geometric configuration of the riser. However, despite the considerable number of published studies on the influence of riser exit geometry on the hydrodynamics in gas-solid circulating fluidized beds, very few have addressed the importance of the riser geometry on the FCC reactor performance.

Due to the complexity of the flow and extreme operating conditions, experimental studies of industrial FCC process are rarely found. In this context, computational fluid dynamic (CFD) tools have been used as a way to better understand these phenomena and look for alternatives to improve the reactor performance.

Despite the complex hydrodynamics present in these reactors, most numerical studies in FCC risers that take into account cracking reactions and related phenomena simplify some important fluid dynamics aspects, like considering that the riser follows a plug flow model [14–18] or applying empirical radial dispersion models to consider the core-annulus patterns [19–21].

Through the simulation of FCC process using three different models (plug flow, one-dimensional, and two-dimensional dispersion models), Deng et al. [21] showed that the consideration of radial nonuniformity is necessary since it provides a better approximation, reflecting in better results for conversions and yields when compared with experimental plant data. Afterward, Lopes et al. [22] simulated an industrial riser using 3D models and showed that the flow presents nonuniformities and asymmetric patterns which are dependent upon feed flow rates. They also observed that these dynamic characteristics of the flow influence the yields of the cracking reactions. Based on this, the use of simplified models can fail to accurately predict the flow in industrial risers.

Another common simplification applied in numerical studies of FCC processes is related to the instantaneous vaporization of the feedstock, which can result in an incorrect representation of the reactor performance. This assumption is justified in many studies (e.g., [14, 23]) by the rapid vaporization of feed. However, Theologos et al. [24], Gupta and Rao [25], and Lopes et al. [26] presented significant differences in product yields when the feedstock vaporization was modeled using instantaneous or rate vaporization models.

In the present work, different designs of the riser exit are studied using CFD techniques. Sophisticated models which take into account vaporization of the liquid droplets, heterogeneous cracking reactions, and catalyst deactivation are applied to the model in order to simulate the reactive flow in industrial scale FCC riser reactors. A parametric study of the hydrodynamic behaviors induced by use of different exit configurations and their effects on the reactions yields is performed. The results obtained are then used to develop a criterion for specifying the riser exit geometries which can improve the reactor efficiency.

MATHEMATICAL MODEL

The three-phase model used in this work considers a three-dimensional gas-liquid-solid flow including heat transfer, droplets vaporization, and chemical reactions. This model uses an Eulerian description of the gas and solid phases, while the Lagrangian approach was chosen to describe the liquid droplets. This phase consists of droplets represented as spherical particles and dispersed in the continuous phase. Details of these models are shown below.

Eulerian Gas-solid Model

In the Eulerian approach both gas and solid phases are modeled as continuous phases. The characteristics of these phases are then determined by solving the mass, momentum, and energy transport equations and applying closure equations to predict the interaction between the phases.

Since the turbulence in the gas phase was modeled using the Reynolds Stress Model (RSM), one transport equation for each of the Reynolds stress tensor components should be solved. As being an anisotropic model, the RSM is able to predict the flow characteristics in regions where sudden changes in flow directions are expected.

The momentum transfer was modeled using the Gidaspow drag model, which combines the Wen Yu correlation with the Ergun equation. The heat transfer was predicted using the Ranz-Marshall correlation for the Nusselt number. Fluctuations in particle velocity were modeled using the Kinetic Theory of Granular Flow. This theory applies the concept of granular temperature, a quantity related to the kinetic energy due to particle movement, to provide closure terms for the solid-phase stress terms. An algebraic formulation, which assumes that the generation and dissipation of the granular temperature are in equilibrium, was applied to estimate this quantity.

A summary of the governing and constitutive equations can be seen in Table 1.

Table 1: Summary of governing and constitutive equations for Eulerian approach

Gas-solid Eulerian flow model		
Governing equations		
Continuity equations:		
Gas-phase:	$\frac{\partial}{\partial t}(\varepsilon_g \rho_g) + \nabla \cdot (\varepsilon_g \rho_g \mathbf{u}_g) = n_d \frac{\partial m_d}{\partial t}$	(A)
Solid-phase:	$\frac{\partial}{\partial t}(\varepsilon_s \rho_s) + \nabla \cdot (\varepsilon_s \rho_s \mathbf{u}_s) = 0$	(B)
Momentum equations:		
Gas-phase:	$\frac{\partial}{\partial t}(\varepsilon_g \rho_g \mathbf{u}_g) + \nabla \cdot (\varepsilon_g \rho_g \mathbf{u}_g \mathbf{u}_g) = \nabla \cdot \left[\varepsilon_g \mu_g \left(\nabla \mathbf{u}_g + (\nabla \mathbf{u}_g)^T \right) - \frac{2}{3} \varepsilon_g \mu_g (\nabla \cdot \mathbf{u}_g) \mathbf{I} - \varepsilon_g \rho_g \overline{\mathbf{u}'\mathbf{u}'} \right]$ $+ \varepsilon_g \rho_g \mathbf{g} - \varepsilon_g \nabla p + \beta(\mathbf{u}_s - \mathbf{u}_g) + \mathbf{u}_g n_d \frac{\partial m_d}{\partial t}$	(C)
Solid-phase:	$\frac{\partial}{\partial t}(\varepsilon_s \rho_s \mathbf{u}_s) + \nabla \cdot (\varepsilon_s \rho_s \mathbf{u}_s \mathbf{u}_s) = \nabla \cdot \left[\varepsilon_s \mu_s \left(\nabla \mathbf{u}_s + (\nabla \mathbf{u}_s)^T \right) - \frac{2}{3} \varepsilon_s \mu_s (\nabla \cdot \mathbf{u}_s) \mathbf{I} \right] + \varepsilon_s \rho_s \mathbf{g}$ $- \varepsilon_s \nabla p - \nabla p_s + \beta(\mathbf{u}_g - \mathbf{u}_s)$	(D)
Energy equations:		
Gas-phase:	$\frac{\partial}{\partial t}(\varepsilon_g \rho_g H_g) + \nabla \cdot (\varepsilon_g \rho_g \mathbf{u}_g H_g) = \nabla \cdot (\varepsilon_g \lambda_g \nabla T_g) + \gamma(T_s - T_g) + \sum_r \Delta H_r \frac{\partial(\varepsilon_g \rho_g Y_{g,i})}{\partial t}$ $+ H_g n_d \frac{\partial m_d}{\partial t}$	(E)

Solid-phase:	$\dfrac{\partial}{\partial t}(\varepsilon_s \rho_s H_s) + \nabla \cdot (\varepsilon_s \rho_s \mathbf{u}_s H_s) = \nabla \cdot (\varepsilon_s \lambda_s \nabla T_s) + \gamma(T_g - T_s)$	(F)
Species conservation:	$\dfrac{\partial t}{9}(\varepsilon^{\partial} b^{\partial} \lambda^{\partial\eta}) + \triangle \cdot (\varepsilon^{\partial} b^{\partial} \mathbf{n}^{\partial} \lambda^{\partial\eta}) = \triangle \cdot (\varepsilon^{\partial} b^{\partial} L^t \triangle \lambda^{\partial\eta}) + \underline{\mathbb{K}}^t + \lambda^{\partial\eta} \mu^{\eta} \dfrac{\partial t}{9\mu\eta}$	(G)
Reynolds stresses:	$\dfrac{\partial}{\partial t}(\varepsilon_g \rho_g \overline{\mathbf{u}'\mathbf{u}'}) + \nabla \cdot (\varepsilon_g \mathbf{u} \rho_g \overline{\mathbf{u}'\mathbf{u}'}) = -\varepsilon_g \rho_g \left[\overline{\mathbf{u}'\mathbf{u}'}(\nabla\mathbf{u})^T + (\nabla\mathbf{u})\overline{\mathbf{u}'\mathbf{u}'} \right]$ $+ \nabla \cdot \left[\varepsilon_g \left(\mu_g + \rho_g \dfrac{C_\mu}{\sigma_k} \dfrac{k^2}{\epsilon} \right) \nabla \overline{\mathbf{u}'\mathbf{u}'} \right]$ $+ \varepsilon_g \Phi - \dfrac{2}{3} \varepsilon_g \delta \rho_g \epsilon + \Pi_{R,ij}$	(H)
Constitutive equations		
Gas-solid momentum exchange (Gidaspow drag model):		
when $\varepsilon_s > 0.2$,	$\beta = 150 \dfrac{\varepsilon_s^2 \mu_g}{\varepsilon_g d_s^2} + \dfrac{7}{4} \dfrac{|\mathbf{u}_s - \mathbf{u}_g| \varepsilon_s \rho_g}{d_s}$	(I)
when $\varepsilon_s \leq 0.2$,	$\beta = \dfrac{3}{4} C_D \dfrac{|\mathbf{u}_s - \mathbf{u}_g| \varepsilon_s \rho_g}{d_s} \varepsilon_s^{-2.65}$	(J)
for Re > 1000,	$C_D = 0.44$	(K)
for Re < 1000	$C_D = \dfrac{24}{Re}(1 + 0.15 Re^{0.687})$	(L)
Gas-solid heat exchange:	$\gamma = \dfrac{(2 + 0.6 Re^{0.5} Pr^{0.3}) 6 \lambda \varepsilon_s \varepsilon_g}{d_s^2}$	(M)
Kinetic theory of granular flow:		
Solids pressure:	$p_s = \varepsilon_s \rho_s \Theta_s + 2\rho_s(1 + e_{ss}) \varepsilon_s^2 g_{0,ss} \Theta_s$	(N)
Radial distribution:	$g_{0,ss} = \left[1 - \left(\dfrac{\varepsilon_s}{\varepsilon_{s,max}} \right)^{1/3} \right]^{-1}$	(O)
Granular temperature:	$\Theta_s = \left\{ \dfrac{-K_1 \varepsilon_s \nabla \cdot \mathbf{u}_s + \sqrt{(K_1 \varepsilon_s \nabla \cdot \mathbf{u}_s)^2 + 4K_4 \varepsilon_s \left[K_2 (\nabla \cdot \mathbf{u}_s)^2 + 2K_3 (\nabla \cdot \mathbf{u}_s^2) \right]}}{2\varepsilon_s K_4} \right\}^2$	(P)

where,		
	$K_1 = 2\left(1 + e_{ss}\right)\rho_s g_{0,ss}$	(Q)
	$K_2 = \dfrac{4d_s\rho_s\left(1 - e_{ss}\right)\varepsilon_s g_{0,ss}}{3\sqrt{\pi}} - \dfrac{2}{3}K_3$	(R)
	$K_3 = \dfrac{d_s\rho_s}{2}\left\{\dfrac{\sqrt{\pi}}{3\left(3 - e_{ss}\right)}\left[1 + 0.4\left(1 - e_{ss}\right)\left(3e_{ss} - 1\right)\varepsilon_s g_{0,ss}\right] + \dfrac{8\varepsilon_s g_{0,ss}\left(1 + e_{ss}\right)}{5\sqrt{\pi}}\right\}$	(S)
	$K_4 = \dfrac{12\left(1 - e_{ss}^2\right)\rho_s g_{0,ss}}{d_s\sqrt{\pi}}$	(T)

Catalytic Cracking Kinetic Model

Many complex reactions occur simultaneously inside the reactor during the FCC process. In order to simplify this kinetic net, a technique widely used is to describe the complex mixtures of hydrocarbons as a reduced number of component groups or lumps. This produces a small number of representative pseudocomponents reacting with each other. This work used a four-lump model proposed by lee et al. [29], in which species with similar properties are grouped into four different lumps: gas oil, gasoline, light gases, and coke. Each lump was defined according to the number of carbons in the molecules as presented in Table 2. The representative reactions of this kinetic scheme are shown in Figure 1.

Table 2: Definition of lumps [27]

Lump	Number of carbons
Gas oil	C_{13} and higher
Gasoline	$C_5 - C_{12}$

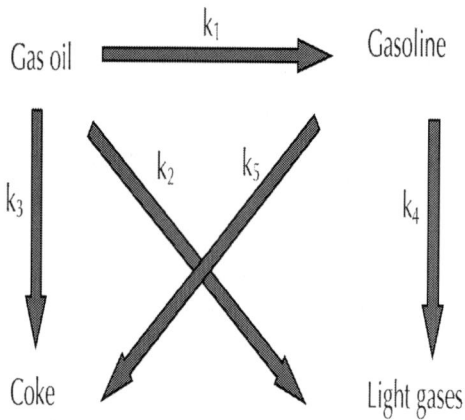

Figure 1: Kinetic scheme of the four-lump reaction model.

The general rate equation for reaction r is given by

$$R_{i,r} = k_r C_i^2 \phi.$$

(1)

The dependence of kinetic constants on temperature is given by the Arrhenius equation:

$$k_r = k_r^0 \exp\left(-\frac{E_r}{RT}\right).$$

(2)

The parameter ϕ, appearing in (1), is the catalyst activity function which is related to the deposition of coke on the catalyst surface. It is expressed by

$$\phi = \exp(-K_c q_1),$$

(3)

where K_c is the activity constant estimated by Farag et al. [27] as a function of the catalyst type. The value obtained for FCC10 catalyst (sample free of metal traps, nickel, and vanadium) was used in this study. The specific coke concentration, q_1, is given by

$$q_1 = \frac{C_{coke}}{\rho_s \varepsilon_s}.$$

(4)

For simplification, it is assumed that coke is not physically deposited on the catalyst surface.

The net source of chemical species i due to reaction (see equation (G) in (Table 1)) is estimated as a sum of the Arrhenius reaction sources over the N_r reactions that the species participate in:

$$\hat{R}_i = M_{w,i} \sum_{r=1}^{N_r} (\nu_i R_{i,r}),$$

(5)

where ν_i is the stoichiometric coefficient of the species i, which is positive for products and negative for reagents.

The values used for the kinetic constants were those obtained by Farag et al. [27] for FCC10 catalyst at the temperature of 823 K. Farag et al. [27] concluded that the overcracking of the gasoline formed was negligible because the kinetic constant values for the cracking of the gasoline obtained in their study were very close to zero. Since they do not estimate the activation energies and the heats of reaction, the values reported by Juárez et al. [30] and Han and Chung [16] were adopted. These values are listed in Table 3.

Table 3: Kinetic constants, activation energies, and heats of reaction

Reaction r	$k_r \mid_{823K}$ $(m^6 kmol^{-1} kg_{cat}^{-1} s^{-1})$	E_r $(Jmol^{-1})$	ΔH_r $(kJkg^{-1})$
Gas oil → gasoline	20.4	57360	195
Gas oil → light gases	7.8	52750	670
Gas oil → coke	3.0	31820	745

The kinetic constants, given in Table 3, were evaluated at 823 K and are dependent on the amount of solids. In order to predict these values at any temperature and catalyst concentration, the preexponential factor was isolated from (2), applied for this temperature, and multiplied by the local concentration of solids. Then the kinetic constants were evaluated as follows:

$$k_r(T, \varepsilon_s) = k_{r,823\,\mathrm{K}}(\rho_s \varepsilon_s) \exp\left[-\frac{E_r}{R}\left(\frac{1}{T} - \frac{1}{823\,\mathrm{K}}\right)\right].$$

(6)

Lagrangian Discrete Phase Model

Heavy oil is injected into the reactor as liquid droplets, which are modeled in this work using a Lagrangian discrete phase model. The Lagrangian approach gives more accurate results than the Eulerian approach, since equations for position, velocity, temperature, and masses of species are solved individually for each discrete particle.

The trajectory of each droplet is predicted solving the forces acting on the particle: the force of gravity and the drag with the gas phase. The Morsi and Alexander drag correlation for spherical particles [31] is used to model the drag coefficient.

According to the heat balance equations, these droplets are heated by the gas phase at higher temperatures, reducing their diameters until the liquid phase is completely vaporized. While the droplet temperature is below its vaporization temperature, convective heat is transferred from the continuous phase to the discrete phase and the droplet temperature is predicted using a heat balance. In this process, the heat transfer coefficient was estimated using the Ranz-Marshall correlation.

In the temperature range between the vaporization and the boiling temperature, mass is transferred from the liquid droplet to the continuous phase. The droplet temperature is then increased according to a heat balance that relates the droplet sensible heat change to the convective and latent heat transfer between the droplet and the continuous phase. When the droplet reaches the boiling temperature, it is assumed to be constant, and the diameter of the droplet begins to decrease until complete vaporization.

Since gas oil droplets rapidly vaporize, particle-particle interactions and effects of the particle volume fraction on the gas phase are negligible. The dispersion of particles due to turbulence in the fluid phase was also neglected.

A summary of the equations applied to model the discrete droplets can be seen in Table 4.

Table 4: Summary of Lagrangian equations

Lagrangian discrete phase model		
Discrete phase trajectory:	$\dfrac{dx_d}{dt} = u_d$	(A)
Discrete phase velocity:	$\dfrac{du_d}{dt} = \dfrac{\mathbf{g}\left(\rho_d - \rho_g\right)}{\rho_d} + F_D\left(u_g - u_d\right)$	(B)
Drag force:	$F_D = \dfrac{18\mu_g}{\rho_d d_d^2}\dfrac{C_D^d \mathrm{Re}_d}{24}$	(C)
Droplet Reynolds number:	$\mathrm{Re}_d = \dfrac{\rho_g d_d \left|u_d - u_g\right|}{\mu_g}$	(D)
Drag coefficient:	$C_D^d = a_1 + \dfrac{a_2}{\mathrm{Re}} + \dfrac{a_3}{\mathrm{Re}^2}$	(E)
Heat balances:		
$T_d < T_{vap}$:	$m_d C_p \dfrac{dT_d}{dt} = hA_d(T_\infty - T_d)$	(F)
$T_{vap} \leq T_d < T_{bp}$:	$m_d C_p \dfrac{dT_d}{dt} = hA_d(T_\infty - T_d) + \dfrac{dm_d}{dt}L$	(G)
Heat transfer coefficient:	$h = \dfrac{\lambda}{d_d}(2.0 + 0.6\mathrm{Re}_d^{1/2}\mathrm{Pr}^{1/3})$	(H)

Mass transfer:	$$\frac{dm_d}{dt} = -k_c(C_{i,\text{surf}} - C_{i,\infty})A_d M_{w,i}$$	(I)
Mass transfer coefficient:	$$k_c = \frac{D_{m,i}}{d_d}(2.0 + 0.6\text{Re}_d^{1/2}\text{Sc}^{1/3})$$	(J)
Droplet diameter variation:	$$-\frac{d(d_d)}{dt} = \frac{4k_\infty(1 + 0.23\sqrt{\text{Re}_d})}{\rho_d C_{p,\infty}d_d}\ln\left[1 + \frac{(T_\infty - T_d)}{L}\right]$$	(K)

SIMULATIONS

The commercial code ANSYS FLUENT 12.0 was used to solve the proposed model. Appropriate user-defined functions were developed to implement the heterogeneous kinetics and the catalyst deactivation model into the software. FLUENT applies the finite volume method, where the domain is divided into a finite number of control volumes in which discrete variables are calculated. The pressure field was determined using a pressure-based approach in which the continuity and momentum equations are manipulated to obtain an approximate equation for the pressure. The least squares cell-based method was applied to evaluate the diffusion and pressure gradients. The convective terms were discretized using a first-order upwind difference scheme, and transient terms were approximated using a first-order implicit scheme. The set of algebraic approximate equations was solved in a segregated way, using the simple algorithm for the pressure-velocity coupling [32].

The geometry of the simulated riser was proposed based on configurations found in the literature. The height and diameter of the column were based on an industrial riser reported by Ali et al. [14] and are shown in Table6. Since they do not report details about the entrance and the outlet geometric configurations of that riser, a lateral entrance located 2 m from the base of the riser was adopted, for the catalyst particles feeding. Twelve 0.5 in diameter ducts at a 30° angle to the main duct were used to feed in the liquid droplets. These configurations are commonly found in studies of

CFB risers [33–35] and about the vaporization of the FCC feedstock [24, 36,37]. This riser reactor geometry is illustrated in Figure 2.

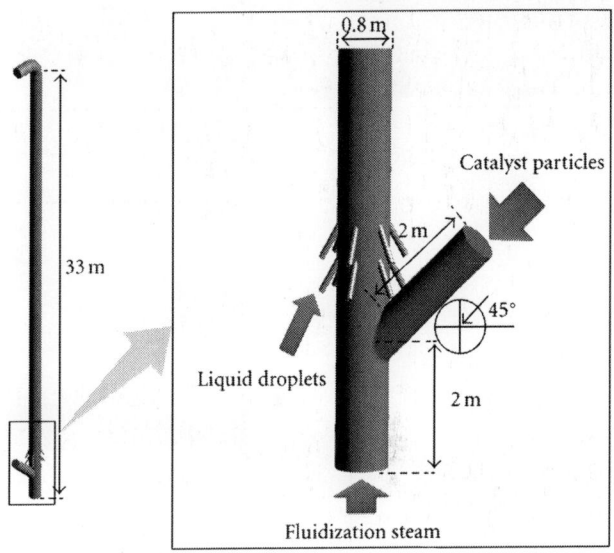

Figure 2: Geometry of the industrial riser.

Four different kinds of configurations for the riser outlet (Figure 3) are proposed in the present study, based on the works of Cheng et al. [1], Gupta and Berruti [2], Van Der Meer et al. [3], Harris et al. [4], Chan et al. [5], Pugsley et al. [9], and Van Engelandt et al. [10]. These configurations can be classified in two main groups: abrupt exits, in which there are sharp changes in the flow direction, inducing solids recirculation in the top section of the riser, and smooth exits, in which the flow changes according to the outlet bend, following a natural path.

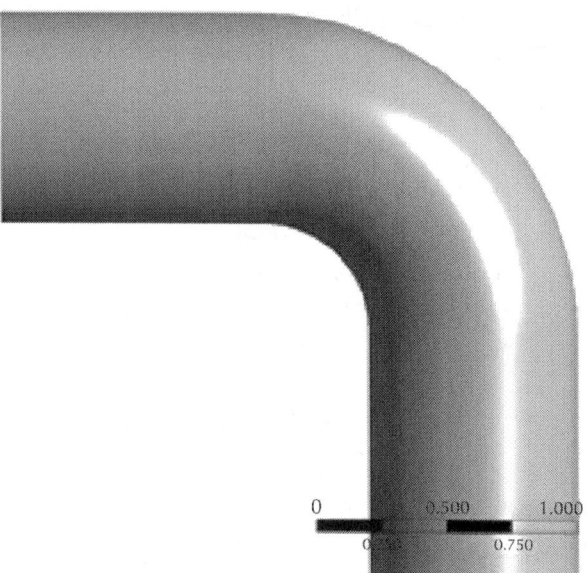

(a)

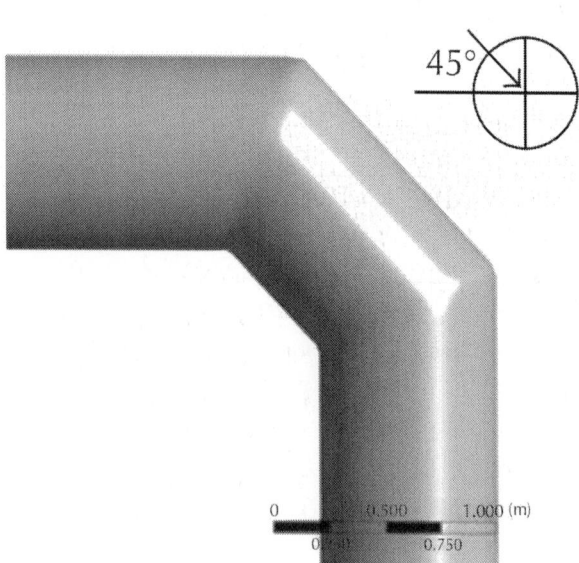

(b)

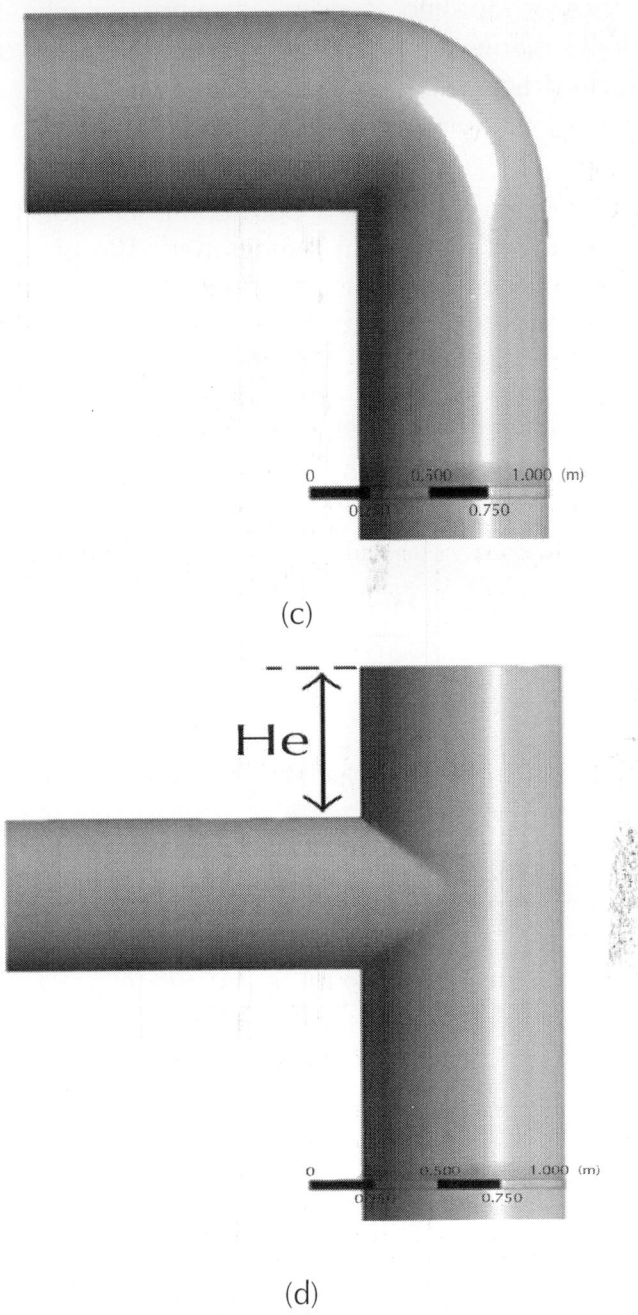

(c)

(d)

Figure 3: Exit configurations used in the simulations.

The meshes applied consist of approximately 1.2 million hexahedral volumes depending on the configuration used for the riser exit. In order to guarantee that smaller control volumes are present where variable gradients are steeper, nonuniform grids are used.

Since the heavy oil is not a pure component, its vaporization does not occur in a constant temperature. The properties of the liquid heavy oil and the values used for the range of temperature in which its vaporization occurs were taken from Nayak et al. [17] and are shown in Table 5.

Table 5: Physical properties of the liquid feed oil

Density	Vaporization temperature	Boiling temperature	Specific heat	Thermal conductivity	Latent heat
870 kg m^{-3}	530 K	560 K	1040 J kg^{-1} K^{-1}	0.2 W m^{-1} K^{-1}	4 × 10^5 J kg^{-1}

Table 6: Geometric details and operating conditions of the industrial riser

Riser height (m)	33
Riser diameter (m)	0.8
Mass flux of feed oil (kg m^{-2} s^{-1})	40
Catalyst-to-oil ratio (kg$_{cat}$/kg$_{gasoil}$)	7
Steam (wt%)	7
Feed oil temperature (K)	500
Catalyst inlet temperature (K)	960
Droplet inlet diameter (mm)	100

The operating conditions used in the simulations are the same of the industrial riser reported by Ali et al. [14]. They are listed in Table 6. About 7% of the total steam is fed with the catalyst particles at the lateral entrance, while the remaining steam is injected at the base of the reactor to help fluidization inside the riser. The physical properties and characteristics of the reactive species and the catalyst were taken from Martignoni and De Lasa [38] Van Landeghem et al. [19] and are listed in Table 7. The density of each reactive species was considered constant, independent of the system pressure and temperature. However, the density of the gas phase varies as the mixture composition changes with cracking reactions.

Table 7: Physical properties of reactive species and catalyst

Species	Vaporized gas oil	Gasoline	Light gases	Coke	Catalyst
Density (kg m^{-3})	6	1.5	0.8	1400	1400
Specific heat (J kg^{-1} K^{-1})	2420	2420	2420	1090	1090
Thermal conductivity (W m^{-1} K^{-1})	0.025	0.025	0.025	0.045	0.045
Molecular weight (kg kmol^{-1})	400	100	50	400	—
Particle diameter (mm)	—	—	—	—	65
Viscosity (kg m^{-1} s^{-1})	5.0 × 10^{-5}	1.6 × 10^{-5}	1.6 × 10^{-5}	1.7 × 10^{-5}	1.7 ×10^{-5}

The simulations were carried out using a parallel code with 16 partitions on computers provided with Intel Xeon 3 GHz quad-core processors. The time step used for the solution was 10^{-3} s. The convergence criterion for advancing in time was that the RMS residuals were less than 10^{-4}. Initially 2 s of transient flow, in which just the steam and the catalyst particles were present, were calculated. After this time, the feed oil droplets started to be injected into the reactor and 8 s more of flow were simulated. Then the time average values of the flow variables were obtained for five additional seconds. About one day of calculation was necessary to predict each second of reactive flow for each simulation.

RESULTS AND DISCUSSION

In the FCC process, heavy oil is injected into the riser reactor as liquid droplets. The cracking reactions are just initiated when these droplets vaporize and the reactive species are transferred from the liquid to the gas phase. Many studies about the FCC reactors, assume instantaneous vaporization of the oil. Lopes et al. [26], however, observed a significant difference in product yields when the hypothesis of instantaneous vaporization was adopted and when this phenomenon is modeled. Based on these results, the vaporization of the droplets was taken into account in the present work. The diameter of the droplets and their residence time inside the reactor are shown in Figure 4. These results were collected 10 s after the start of the liquid injection. At this moment, the complete vaporization of the liquid droplets takes place at a height of approximately 9 m.

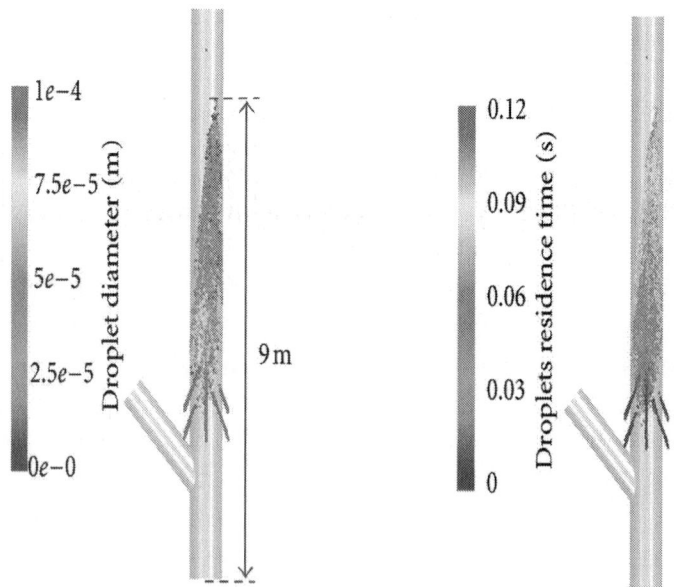

Figure 4: Droplets' diameter and residence time inside the reactor.

The maximum residence time of the droplets before they are completely vaporized was monitored along the time. As can be seen in Figure 5, the maximum value observed was 0.18 s, obtained 1.25 s after the start of the liquid droplets injection. After this time, the maximum residence time decreases, reaching an average value of 0.12 s. According to Liu and Han [39] (cited in [37]) the complete vaporization of the charge is completed within 0.2 s, in risers of typical FCC units. This shows that the model used in the simulations is appropriate to describe the phenomenon of vaporization. The number of liquid droplets present in the system follows the same trend (Figure 6), reaching its maximum value about two seconds after the beginning of the charge injection into the reactor. The average number of liquid droplets inside the riser reactor is about 120 thousand. These results emphasize the importance of simulating the vaporization of the liquid oil charge stock in the FCC process.

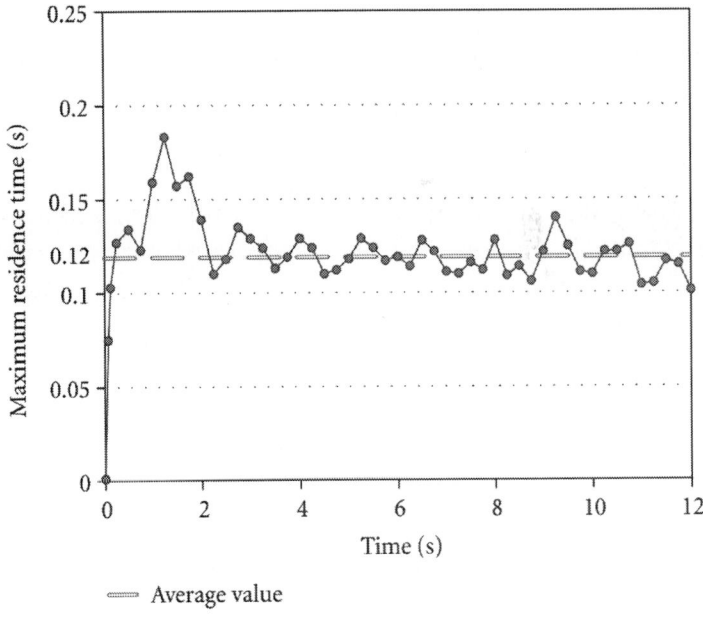

Figure 5: Maximum residence time of the droplets.

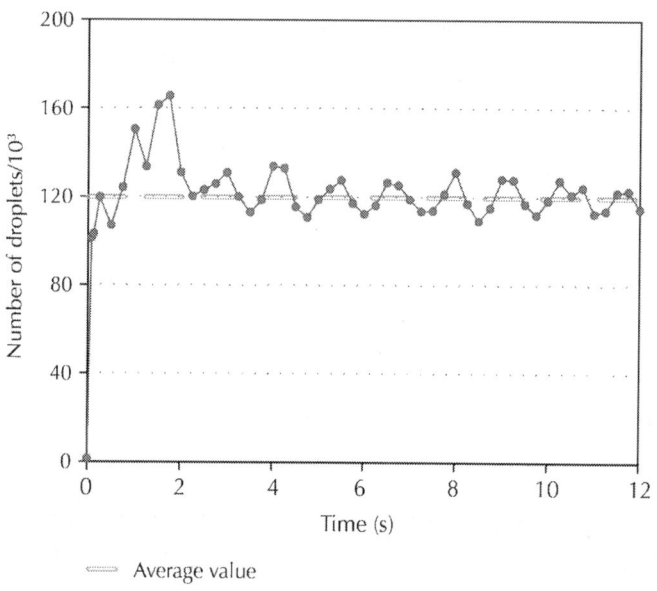

Figure 6: Instantaneous number of droplets.

The inlet zone of the riser is a very complex part of the reactor, where intense turbulence and flow inhomogeneities are observed. A detailed study about this region was performed by Lopes et al. [22], where it is shown that, depending on the operating conditions applied, the inlets arrangement can affect the flow patterns along all riser height. Extremely complex hydrodynamics has been also observed in the outlet region of CFB risers [3, 9, 10], which can affect the reactor efficiency.

In order to verify the influence of different riser exit configurations on reactor performance, five cases were initially proposed. The four geometric configurations shown in Figure 3 were tested. Two variations of the abrupt T exit (Figure 3(d)) with different projected heights (He) were also simulated. The definition of each case according to the configuration applied in the simulations can be seen in Table 8.

Table 8: Definition of simulated cases

Case	Configuration[1]	He (m)[1]
1	A	—
2	B	—
3	C	—
4	D	0
5	D	0.8

[1]According to Figure 3.

The effect of the exits on the flow was evaluated using parameters reported by Van Der Meer et al. [3] and Schut et al. [28]. Their definitions are listed in Table 9. These parameters help to quantify some important characteristics of the flow in the riser, such as core-annulus patterns, the increase in the downflux in abrupt exit risers, and the asymmetry of the flow.

Table 9: Parameters applied to evaluate the influence of the exits on the flow

Parameter	Definition
H_{bf}	Backflux relative height: distance from the top of the riser traveled by the particulate phase in downward motion divided by riser height.
k_a	Exit reflection coefficient: fraction of the riser transversal area within which solids move downwards.
K_a^*	Asymmetry of k_a: defined by the difference between the fraction of the area with negative velocity of solids located at the positive and the negative side of the x-axis
δ	Film thickness of downflow: given by $D\left(1-\sqrt{1-k_a}\right)/2$ [28].

The values of parameter H_{bf} obtained for each of the five cases are shown in Table 10. For Cases 1 and 2 these values are zero, indicating that under the conditions applied, there is no solids reflux when configurations A and B (Figure 3) are used. These results help to characterize these configurations as smooth exits, since they do not restrict the solids flux. Despite the similarity between configurations A and C (Figure 3), the case simulated using the latter (Case 3) provided some solids backmixing ($H_{bf} = 0.036$) induced by the sharp right angle in this configuration. The maximum reflux of particles was obtained when the T-shape exit with a projected height of 0.8 m was used (Case 5). In this case, solids flowing in counter current to the main flow reached a distance of 9.75 m from the top of the riser, corresponding to about 28% of its total height.

Table 10: H_{bf} obtained for each case

Case	H_{bf}
1	0.000
2	0.000
3	0.036
4	0.117
5	0.279

The movement of solid particles in center planes at the outlet bends is represented as velocity vectors in Figure7. As shown previously, no solids reflux can be observed in Cases 1 and 2, justifying the zero value for parameter obtained in these cases. The internal sharp right angle seen in Cases 3, 4, and 5 induces the formation of recirculation areas just after the bend. A second vortex is formed at the riser wall opposite the outlet opening in Cases 4 and 5, in which the exit is at sharp right angles also at the outer side of the bend. The higher the downward flow of the particles is, the more intense the restriction on the direct flow is, causing an increase in its velocity in the central region of the riser.

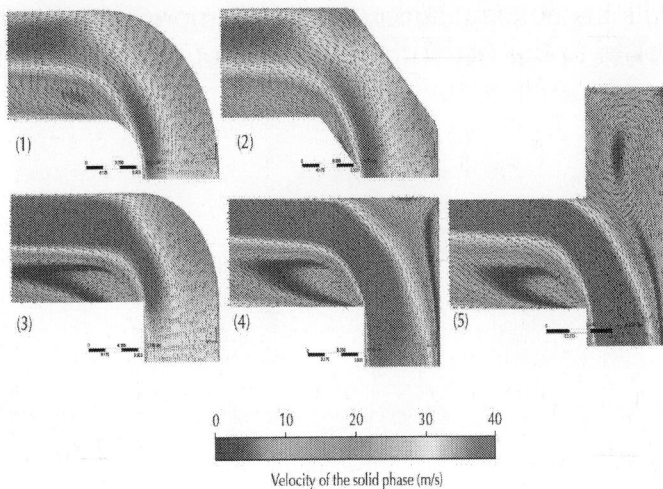

Velocity of the solid phase (m/s)

Figure 7: Solid phase velocity and movement.

Harris et al. [40] described a physical mechanism to help to explain the motion of solids at a riser exit. They postulate that when solids follow a curved path of mean radius R at the exit, the centrifugal acceleration is balanced against the average component of the acceleration due to gravity acting toward the center of the bend. In regions where the solids slow down due to geometric accidents at the exit bend, gravity overlaps centrifugal acceleration, causing a change in trajectory and creating recirculation zones, as observed in the simulated cases.

In Figure 8, solids volume fraction fields are shown for the five simulated cases, where an accumulation of particles at the top of the riser can be seen. As the particle density is much higher than the gas mixture density, the particle inertia is also higher, which means that particles are easily separated from the gas stream near the exit, even at outlets considered smooth. In addition, the abrupt exits create an extra resistance, causing an increase in the concentration of this phase at the outlet bend. As a result, the average solids fraction and the annulus thickness near the top of the riser are much higher in Cases 4 and 5, in which the solids recirculation is also higher. As observed in other studies [10, 11], for risers with abrupt exits the accumulation of solids is more pronounced at the side opposite to the outlet than the side of the riser at which the outlet bend is positioned.

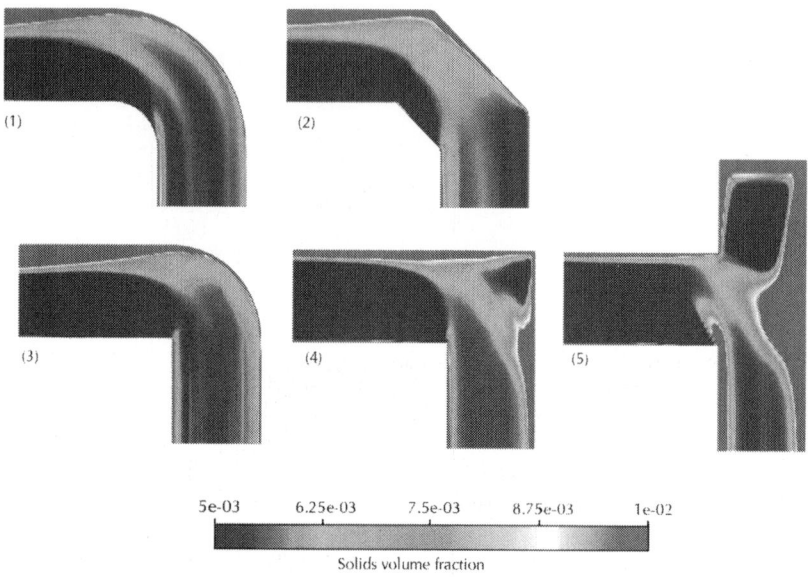

Figure 8: Solids volume fraction profiles at the exits.

The accumulation of solids observed in the cases with abrupt exits results in a densification of the annulus structure near the outlet bends. This is confirmed in Figure 9, which shows the solids volume fraction in radial planes located at heights of 20 and 32 m in Cases 1 and 5. At a height of 20 m the core-annulus patterns are very similar in both cases. However, when approaching the outlet bend there is a coarsening of the annular structure in Case 5 resulting from the solids recirculation observed in this case.

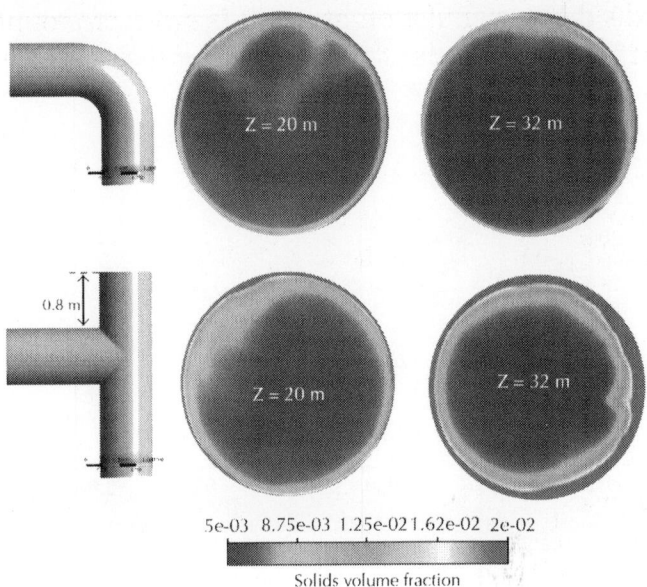

Figure 9: Core-annulus patterns.

In order to analyze the solids accumulation near the exit of the risers with different configurations, average values of the solids volume fraction in transversal planes were taken along the riser height. The shape of the average solids hold-up profile is a good indicator of how the exit effects propagate down the riser. As shown in Figure 10, in all cases studied the average volume fraction initially decreases due to the expansion of the gas phase caused by the catalytic cracking of the large hydrocarbon molecules into smaller ones during the process. The profiles continue decreasing in the cases with smooth exits, but this tendency suddenly change in risers with abrupt exits, creating a profile described as C-shape. The more restrictive the exit configuration is, the further down the riser its effect propagates. This trend is consistent with results reported in many studies on CFB risers [6, 40–42].

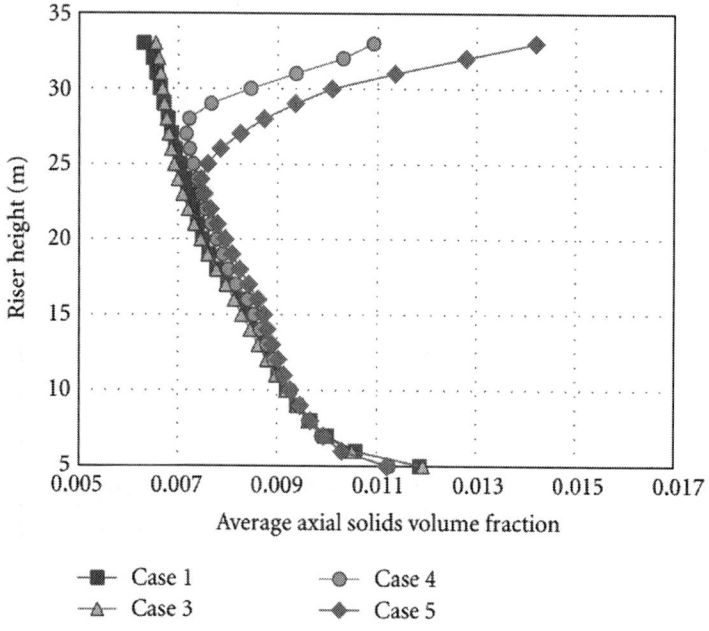

Figure 10: Average solids volume fraction in cross-section planes along the height.

The recirculation of solids observed in the risers with abrupt exits results in an increase in their residence time inside the reactor. As the solid phase is treated in this study as a continuous fluid using Eulerian approach, it is not possible to calculate exactly the residence time of each particle in the system. In order to estimate an approximated time, a methodology applied by liu and Tilton [43], in which average values are estimated, was applied. Initially, a new property (Ψ), with unit of second, was defined for the solid phase. An user-defined scalar (UDS) transport equations was then introduced into the model in the form [32]

$$\frac{\partial}{\partial t}\left(\varepsilon_s \rho_s \Psi\right) + \nabla \cdot \left(\varepsilon_s \rho_s \mathbf{u}_s \Psi\right) = \nabla \cdot \left(\varepsilon_s \Gamma^\Psi \nabla \Psi\right) + S^\Psi. \qquad (7)$$

If the diffusive terms were neglected, the scalar just propagates with the time and the movement of the solid phase. The source term was then implemented via user defined-function (UDF) as

$$S^{\Psi} = \varepsilon_s \rho_s. \qquad (8)$$

As the solution advances in time, this value is added to the scalar. Thus, assuming that this property is equal to zero at the solids entrance and in the problem initialization, a cumulative function that propagates with the time and the solids movement is created. This enables an estimation of an approximated residence time of the solid particles. In Table 11 the solids mean residence time obtained for Cases 1 and 5 is shown. When the abrupt T-shape exit with a projected height of 0.8 m is applied in the FCC riser simulation, the mean residence time is increased by about 18% in relation to the value obtained for the smoother exit. This difference can influence the reaction time, increasing the conversion rates.

Table 11: Residence time

Case	Mean value (s)	
1	1.943	
5	2.299	

The gasoline and coke yields at the riser outlet obtained in the simulated cases are shown in Figure 11. This value increases as resistance to the flow becomes more intense and the solids recirculation increases. Comparing Cases 1 and 5, there is an increase of about 4% in the gasoline yield.

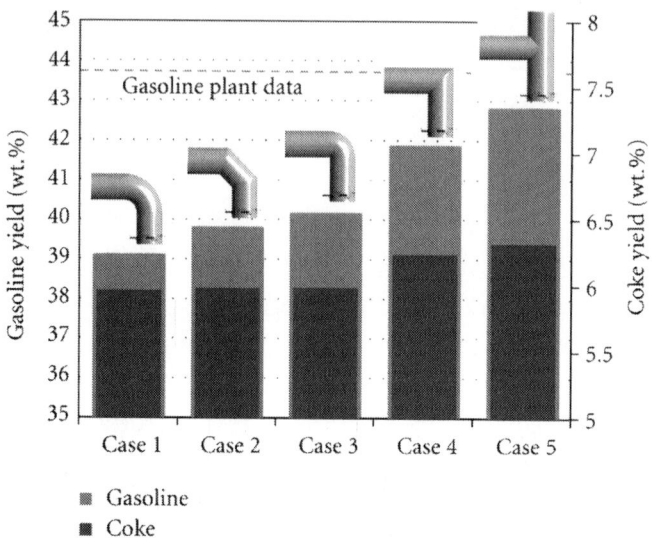

Figure 11: Gasoline and coke yields at the riser outlet.

In order to validate the simulated results with the data reported by Ali et al. [14], the operating conditions and the equipment dimensions used were the same as those of their industrial riser. Since Ali et al. [14] do not report details of the exit configuration of this riser, several configurations were tested in the simulations. However, it is known that the T-shape design, as used in the simulation of Case 5, is the most common configuration found in industry, since it reduces the erosive particle impingement on the roof of the riser. As can be seen in Figure 11, the value of the gasoline yield which most closely approximates the industrial value ((43.9%) was the one obtained in the simulation of Case 5, with a deviation of only about 2%.

Average values of gasoline yield were also taken in transversal planes along the riser height in Cases 1 and 5, as shown in Figure 12. Similarly to the profiles of average solids volume fraction (Figure 10), the gasoline yield profiles coincide up to a certain height, at which point the tendency in the riser changes with the abrupt T-shape exit.

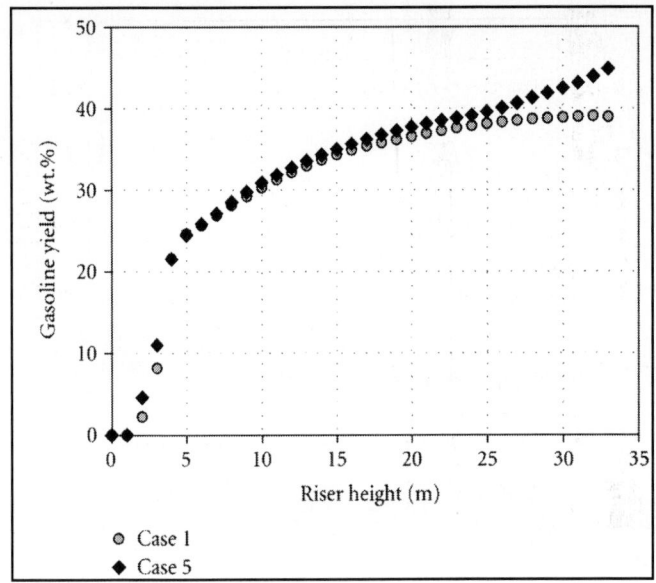

Figure 12: Average gasoline yield in cross-section planes along the height.

The results obtained for the cracked hydrocarbons conversion in the simulation of Case 5 were also compared with plant data sampled by Martin et al. [41] at 4 m above the upper feedstock injection point at two locations along the riser radius (Figure 13). Although they do not provide information on the geometry and operating conditions of the industrial plant for which the data were taken, the comparison of simulated and experimental local cracking conversion showed good agreement. It can be clearly seen that the conversion is higher toward the wall. This can be attributed to the larger catalyst concentration, smaller velocity and higher temperature near the wall, and to the catalyst/gas backmixing at this region. The mass fractions of coke and gasoline in the outlet region for the Cases 1 and 5 are shown in Figure 14, where noticed higher values can be in the zones in which higher recirculation of solids was found (Figure 8).

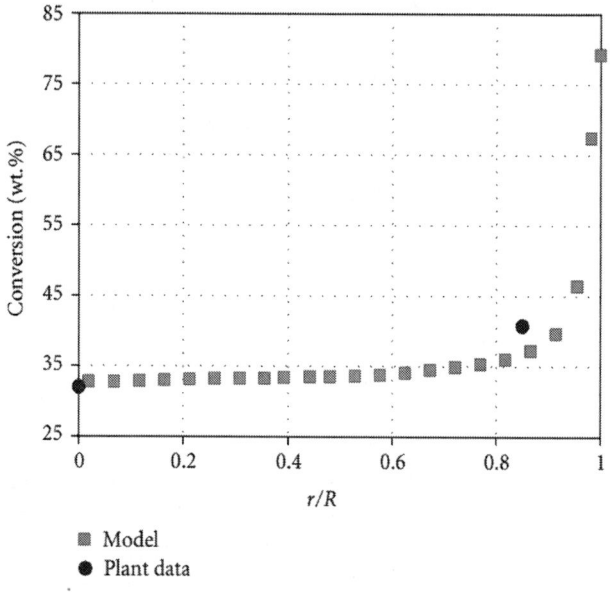

Figure 13: Radial profile of cracking conversion.

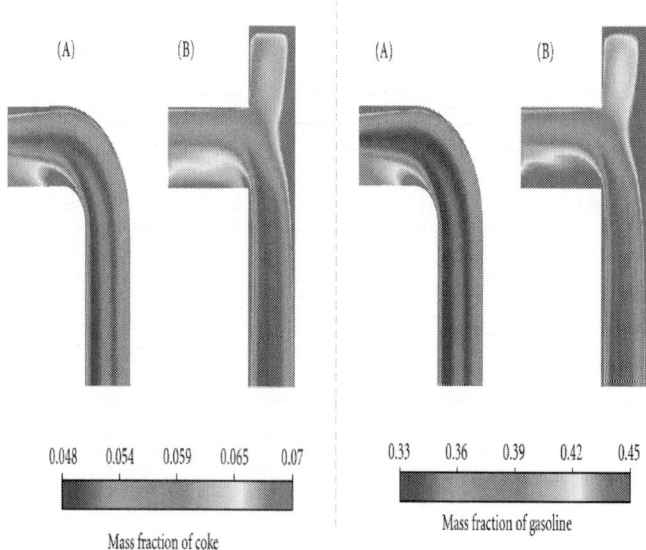

Figure 14: Mass fractions of coke and gasoline for Cases 1 (A) and 5 (B).

As shown above, the geometries applied in the simulation of Cases 1, 2, and 3 (configurations A, B and C in Figure 3) have very small or null reverse flow of solids and therefore can be characterized as smooth exits for the operating conditions and feedstock used in the simulations. In the cases in which the type D configuration with different projected heights is used, there is significant solids recirculation which affects the flow even far from the outlet bend, and therefore it can be classified as abrupt exits. In order to determine the response of the flow to the variation in the projected height of configuration D, the simulation of four new cases, in which He was assumed to be 0.2 m, 0.4 m, 0.6 m, and 1.0 m, was proposed.

The relative height of backflux (represented by parameter H_{bf}) was estimated for each new case, as can be seen in Figure 15. As He increases, there is initially a sharp increase in the value of H_{bf}, which reaches maximum value at He = 0.4 m, where the solids downward flow is detected at a distance from the outlet bend corresponding to about 35% of the length of the riser. Above He = 0.4 m, the value of H_{bf} decreases, tending to stabilize at about 0.25.

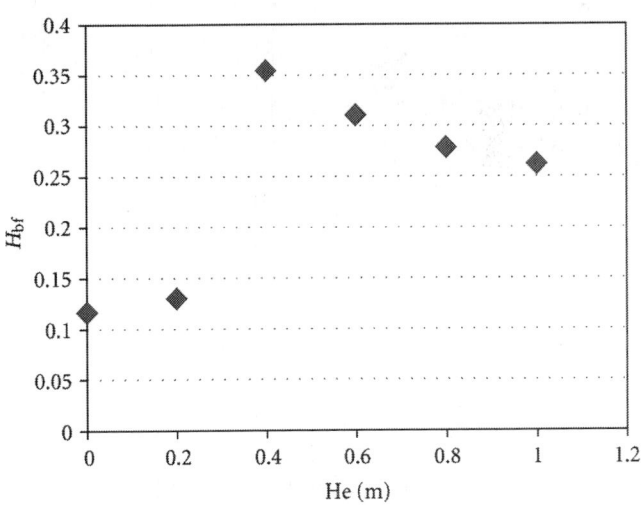

Figure 15: H_{bf} parameter obtained for different projected heights.

The parameters which represent the fraction of crosssectional area of the riser in which there is a downward flow of solids (k_a) and the asymmetry of the flow (K_a^*) were also estimated for each of these cases. They are shown in Figures 16 and 17 according to riser height. There is an increase in the value of ka along the height of the reactor. As shown in Figure 10, the solids accumulation in risers with abrupt exits leads to the enlargement of the annular structure, consequently increasing the fraction of the area occupied by the downward flowing solids. The configuration with He = 0.4 m has the highest values of k_a even below a height of 24 m, which indicates that the solids recirculation is more intense for this case, corroborating the results shown in Figure 15.

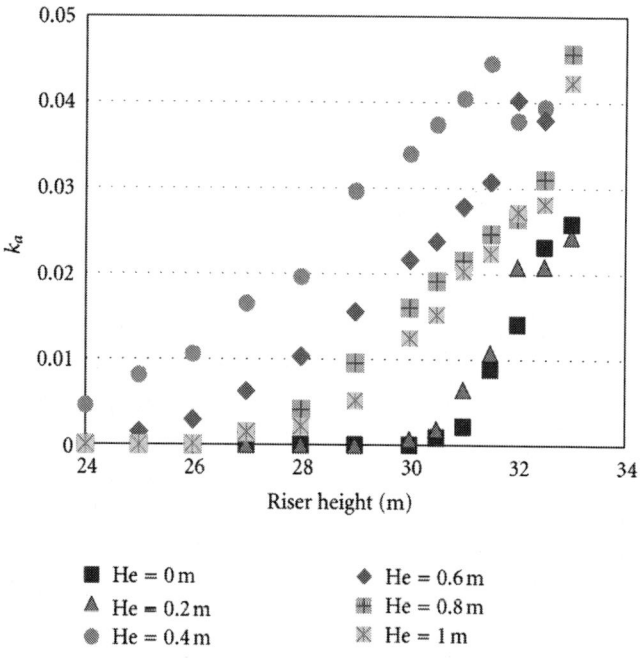

Figure 16: k_a parameter obtained along the height for different projected heights.

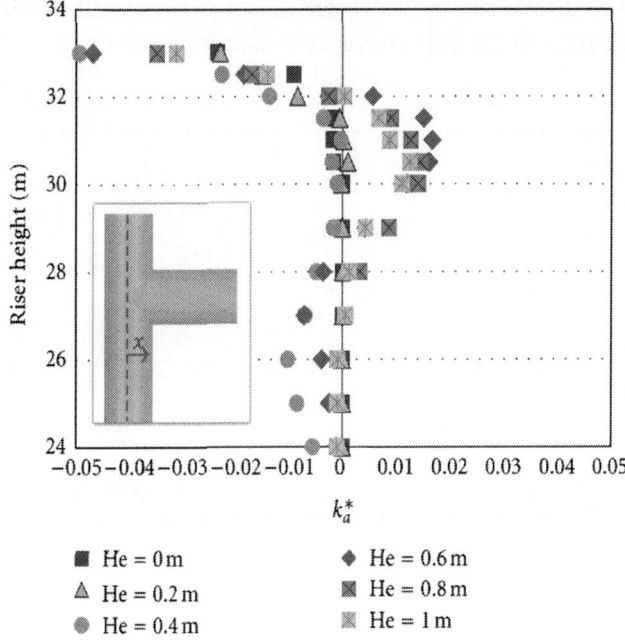

Figure 17: K_a^* parameter obtained along the height for different projected heights.

Analysis of parameter K_a^* (Figure 17) shows that the flow is more asymmetric in regions far from the outlet bend for the exit with He = 0.4 m. Moreover, the displacement of this parameter to the negative side of axis x above a height of 32 m in all cases confirms the patterns seen in Figure 8, in which there is a region of solids accumulation on the side opposite to the outlet bend when abrupt exits are used. These asymmetric patterns near the inner top side wall were also observed by Wang et al. [44] in an experimental study of gassolid circulating fluidized bed using Electrical Capacitance Volume Tomography (ECVT).

The film thickness of the downflow (δ) is indicated in Figure 18, which shows the values of R − δ along the riser height. In all cases, this film is thickest immediately below the outlet bend at a height of 33 m and decreases downward from there. Because of the high solids reflux rate found in the riser with a projected height of 0.4 m, the values found for δ are also higher in this case.

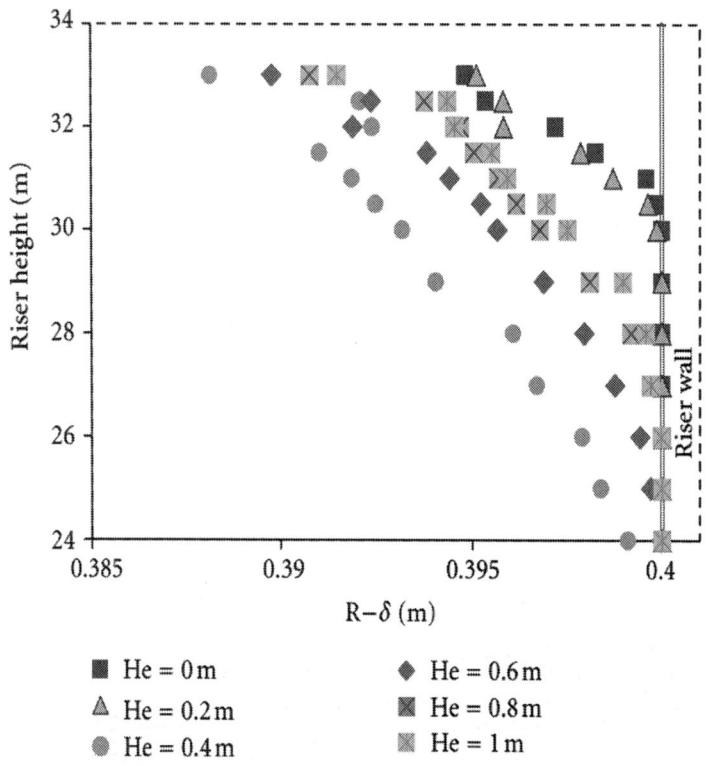

Figure 18: Film thickness of downflow.

In Figure 19 the average values for gasoline yield at the riser outlet obtained for the risers with different projected height are shown. These results follow almost the same tendency as that observed in Figure 15, indicating that solids recirculation increases the formation of gasoline fractions in these cases.

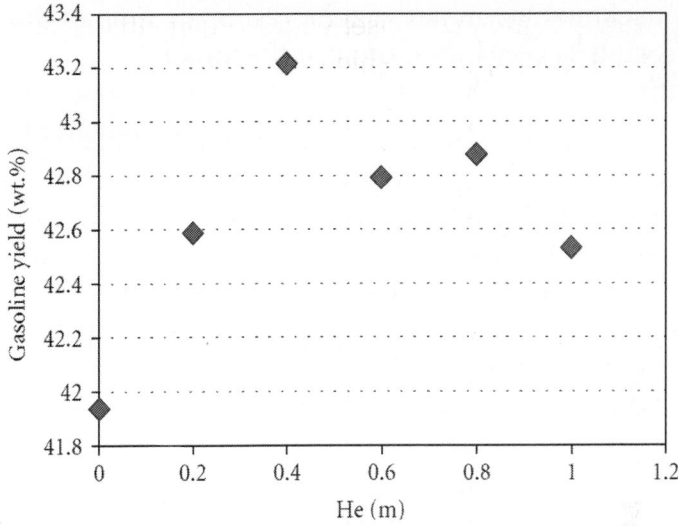

Figure 19: Gasoline yields at the riser outlet.

Under the flux conditions used in the present study, the increase in solids residence time improves the reactor efficiency, mainly in cases with a high solids reflux ratio. In addition, as the catalytic reactions depend on the presence of catalyst in the system and the rate of reaction is linearly dependent on its concentration, the increase in the solids volume fraction at the top section of the riser results in an increase in the reaction rates with higher product yields.

CONCLUSIONS

In the present work, different outlet bend designs for an industrial FCC riser were proposed to study their influence on the dynamic of the flow and consequently on reactor efficiency. A three-dimensional and three-phase reactive flow model was used in order to predict most of the phenomena present in this complex process.

As can be seen from the simulation results, small changes in the geometry of the reactor had a significant effect on the flow patterns and therefore on the product yields. The use of abrupt exits results in solids backmixing, enhancing the residence time of the catalyst and increasing its concentration near the top of the riser. An area of densification is then created near the exit, which, depending on the severity of the exit restriction, extends along the length of the riser.

The increase in the amount of catalyst particles favors the heterogeneous FCC reactions, which are proportional to the solids concentration in the system. As a result, risers with abrupt exits, in addition to reduce the erosive particle impingement on the roof of the equipment, had higher gasoline yields. The T-shape exit with a projected height of 0.4 m gave the highest value for this variable (closer to the industrial data), indicating that this outlet configuration improves the reactor efficiency under the conditions applied in the present study.

It is important to emphasize that computer simulations require detailed information about the feedstock and the catalyst particles as well as the reactor design. This information is rarely found in experimental studies on industrial reactors, making necessary the adoption of some assumptions to make possible their simulation. Nevertheless, the models used and the material properties chosen gave results close to the industrial data obtained under the same operating conditions applied in the simulations, especially when configurations commonly found in industrial plants were applied.

ACKNOWLEDGMENT

The authors gratefully acknowledge the financial support of PETROBRAS for this research.

REFERENCES

1.   Y. Cheng, F. Wei, G. Yang, and J. Yong, "Inlet and outlet effects on flow patterns in gas-solid risers," Powder Technology, vol. 98, no. 2, pp. 151–156, 1998.

2. S. K. Gupta and F. Berruti, "Evaluation of the gas-solid suspension density in CFB risers with exit effects," Powder Technology, vol. 108, no. 1, pp. 21–31, 2000.·

3. E. H. Van Der Meer, R. B. Thorpe, and J. F. Davidson, "Flow patterns in the square cross-section riser of a circulating fluidised bed and the effect of riser exit design," Chemical Engineering Science, vol. 55, no. 19, pp. 4079–4099, 2000.

4. A. T. Harris, J. F. Davidson, and R. B. Thorpe, "The influence of the riser exit on the particle residence time distribution in a circulating fluidised bed riser," Chemical Engineering Science, vol. 58, no. 16, pp. 3669–3680, 2003.

5. C. W. Chan, A. Brems, S. Mahmoudi et al., "PEPT study of particle motion for different riser exit geometries," Particuology, vol. 8, no. 6, pp. 623–630, 2010. ·

6. X. Wang, L. Liao, B. Fan et al., "Experimental validation of the gas-solid flow in the CFB riser," Fuel Processing Technology, vol. 91, no. 8, pp. 927–933, 2010. ·

7. M. T. Lim, S. Pang, and J. Nijdam, "Investigation of solids circulation in a cold model of a circulating fluidized bed," Powder Technology, vol. 226, pp. 57–67, 2012.

8. J. W. Chew, R. Hays, J. G. Findlay et al., "Reverse core-annular flow of Geldart Group B particles in risers," Powder Technology, vol. 221, pp. 1–12, 2012.

9. T. Pugsley, D. Lapointe, B. Hirschberg, and J. Werther, "Exit effects in circulating fluidized bed risers,"Canadian Journal of Chemical Engineering, vol. 75, no. 6, pp. 1001–1010, 1997.

10. G. Van engelandt, G. J. Heynderickx, J. De Wilde, and G. B. Marin, "Experimental and computational study of T- and L-outlet effects in dilute riser flow," Chemical Engineering Science, vol. 66, no. 21, pp. 5024–5044, 2011.

11. J. De Wilde, G. B. Marin, and G. J. Heynderickx, "The effects of abrupt T-outlets in a riser: 3D simulation using the kinetic theory of granular flow," Chemical Engineering Science, vol. 58, no. 3–6, pp. 877–885, 2003.

12. B. Chalermsinsuwan, P. Kuchonthara, and P. Piumsomboon, "Effect of circulating fluidized bed reactor riser geometries on chemical reaction rates by using CFD simulations," Chemical Engineering and Processing, vol. 48, no. 1, pp. 165–177, 2009.

13. A. K. Das, J. De Wilde, G. J. Heynderickx, and G. B. Marin, "CFD simulation of dilute phase gas-solid riser reactors—part II: simultaneous adsorption of SO_2-NOx from flue gases," Chemical Engineering Science, vol. 59, no. 1, pp. 187–200, 2004.

14. H. Ali, S. Rohani, and J. P. Corriou, "Modelling and control of a riser type fluid catalytic cracking (FCC) unit," Chemical Engineering Research and Design, vol. 75, no. 4, pp. 401–412, 1997.

15. I. S. Han and C. B. Chung, "Dynamic modeling and simulation of a fluidized catalytic cracking process—part I: process modeling," Chemical Engineering Science, vol. 56, no. 5, pp. 1951–1971, 2001.

16. I. S. Han and C. B. Chung, "Dynamic modeling and simulation of a fluidized catalytic cracking process—part II: property estimation and simulation," Chemical Engineering Science, vol. 56, no. 5, pp. 1973–1990, 2001.

17. S. V. Nayak, S. L. Joshi, and V. V. Ranade, "Modeling of vaporization and cracking of liquid oil injected in a gas-solid riser," Chemical Engineering Science, vol. 60, no. 22, pp. 6049–6066, 2005.

18. J. S. Ahari, A. Farshi, and K. Forsat, "A mathematical modeling of the riser reactor in industrial FCC unit," Petroleum and Coal, vol. 50, no. 2, pp. 15–24, 2008.

19. F. Van Landeghem, D. Nevicato, I. Pitault et al., "Fluid catalytic cracking: modelling of an industrial riser," Applied Catalysis A, vol. 138, no. 2, pp. 381–405, 1996.·

20. C. Derouin, D. Nevicato, M. Forissier, G. Wild, and J. R. Bernard, "Hydrodynamics of riser units and their impact

on FCC operation," Industrial and Engineering Chemistry Research, vol. 36, no. 11, pp. 4504–4515, 1997.

21. R. Deng, F. Wei, T. Liu, and Y. Jin, "Radial behavior in riser and downer during the FCC process,"Chemical Engineering and Processing, vol. 41, no. 3, pp. 259–266, 2002.

22. G. C. Lopes, L. M. Rosa, M. Mori, J. R. Nunhez, and W. P. Martignoni, "Three-dimensional modeling of fluid catalytic cracking industrial riser flow and reactions," Computers and Chemical Engineering, 2011. ·

23. K. N. Theologos and N. C. Markatos, "Advanced modeling of fluid catalytic cracking riser-type reactors," AIChE Journal, vol. 39, no. 6, pp. 1007–1017, 1993.

24. K. N. Theologos, A. I. Lygeros, and N. C. Markatos, "Feedstock atomization effects on FCC riser reactors selectivity," Chemical Engineering Science, vol. 54, no. 22, pp. 5617–5625, 1999. ·

25. A. Gupta and D. S. Rao, "Model for the performance of a fluid catalytic cracking (FCC) riser reactor: effect of feed atomization," Chemical Engineering Science, vol. 56, no. 15, pp. 4489–4503, 2001.

26. G. C. Lopes, L. M. Da Rosa, M. Mori, J. R. Nunhez, and W. P. Martignoni, "The importance of using three-phase 3-D model in the simulation of industrial FCC risers," Chemical Engineering Transactions, vol. 24, pp. 1417–1422, 2011.

27. H. Farag, A. Blasetti, and H. De Lasa, "Catalytic cracking with FCCT loaded with tin metal traps: adsorption constants for gas oil, gasoline, and light gases," Industrial and Engineering Chemistry Research, vol. 33, no. 12, pp. 3131–3140, 1994.

28. S. B. Schut, E. H. Van Der Meer, J. F. Davidson, and R. B. Thorpe, "Gas-solids flow in the diffuser of a circulating fluidised bed riser," Powder Technology, vol. 111, no. 1-2, pp. 94–103, 2000. ·

29. L. S. Lee, Y. W. Chen, and T. N. Huang, "Four-lump kinetic model for fluid catalytic cracking process,"Canadian Journal of Chemical Engineering, vol. 67, no. 4, pp. 615–619, 1989.
·

30. J. A. Juárez, F. L. Isunza, E. A. Rodrìguez, and J. C. M. Mayorga, "A strategy for kinetic parameter estimation in the fluid catalytic cracking process," Industrial & Engineering Chemistry Research, vol. 36, pp. 5170–5174, 1997.

31. S. A. Morsi and A. J. Alexander, "An investigation of particle trajectories in two-phase flow systems,"The Journal of Fluid Mechanics, vol. 55, no. 2, pp. 193–208, 1972.

32. Ansys Inc. (US), ANSYS FLUENT 12. 0—Theory Guide, Ansys, 2009.

33. W. Zhang, Y. Tung, and F. Johnsson, "Radial voidage profiles in fast fluidized beds of different diameters," Chemical Engineering Science, vol. 46, no. 12, pp. 3045–3052, 1991. ·

34. J. H. Pärssinen and J. X. Zhu, "Particle velocity and flow development in a long and high-flux circulating fluidized bed riser," Chemical Engineering Science, vol. 56, no. 18, pp. 5295–5303, 2001.

35. J. C. S. C. Bastos, L. M. Rosa, M. Mori, F. Marini, and W. P. Martignoni, "Modelling and simulation of a gas-solids dispersion flow in a high-flux circulating fluidized bed (HFCFB) riser," Catalysis Today, vol. 130, no. 2–4, pp. 462–470, 2008.

36. K. N. Theologos, I. D. Nikou, A. I. Lygeros, and N. C. Markatos, "Simulation and design of fluid-catalytic cracking riser-type reactors," Computers and Chemical Engineering, vol. 20, no. 1, pp. S757–S762, 1996.

37. J. Gao, C. Xu, S. Lin, G. Yang, and Y. Guo, "Simulations of gas-liquid-solid 3-phase flow and reaction in FCC riser reactors," AIChE Journal, vol. 47, no. 3, pp. 677–692, 2001.

38. W. Martignoni and H. I. De Lasa, "Heterogeneous reaction model for FCC riser units," Chemical Engineering Science, vol. 56, no. 2, pp. 605–612, 2001.

39. D. L. Liu and J. M. Han, "Evaluation on commercial application of LPC type nozzle for FCC feed,"Process Engineering Resources, vol. 22, article 49, 1992.

40.   A. T. Harris, J. F. Davidson, and R. B. Thorpe, "Influence of exit geometry in circulating fluidized-bed risers," AIChE Journal, vol. 49, no. 1, pp. 52–64, 2003.

41.   M. P. Martin, P. Turlier, J. R. Bernard, and G. Wild, "Gas and solid behavior in cracking circulating fluidized beds," Powder Technology, vol. 70, no. 3, pp. 249–258, 1992.

42.   X. Wu, F. Jiang, X. Xu, and Y. Xiao, "CFD simulation of smooth and T-abrupt exits in circulating fluidized bed risers," Particuology, vol. 8, no. 4, pp. 343–350, 2010.

43.   M. Liu and J. N. Tilton, "Spatial distributions of mean age and higher moments in steady continuous flows," AIChE Journal, vol. 56, no. 10, pp. 2561–2572, 2010.

44.   F. Wang, Q. Marashdeh, A. Wang, and L. Fan, "Electrical capacitance volume tomography imaging of three-dimensional flow structures and solids concentration distributions in a riser and a bend of a gas-solid circulating fluidized bed," Industrial & Engineering Chemistry Research, vol. 51, pp. 10968–10976, 2012.

Parametric Sensitivity Studies in a Commercial FCC Unit

Prabha K. Dasila[1, 2], Indranil Choudhury[3], Deoki Saraf[1], Sawaran Chopra[1], and Ajay Dalai[2]

[1]University of Petroleum & Energy Studies, Dehradun, India
[2]University of Saskatchewan, Saskatoon, Canada
[3]Indian Oil Corporation Ltd., Research & Development Centre, Faridabad, India

ABSTRACT

A steady state model was developed for simulating the performance of an industrial fluid catalytic cracking (FCC) unit which was subsequently used in parametric sensitivity studies. The simulator includes kinetic models for the riser reactor and the regeneration systems. Mass and energy balances were performed for each of these sections and simulation results were compared with the plant data available in the literature. Model predictions were found to be in close agreement with the reported data. Finally this validated model was used for studying the effects of independent variables such as feed preheat temperature (T_{feed}) and feed flow rate (F_{feed}) on the unit performance at either fixed regenerated catalyst temp/ regenerator temp (T_{rgn}) or constant reactor outlet temperature (ROT). The catalyst circulation rate (CCR) was automatically adjusted to keep the ROT constant with varying the independent variables feed preheat temperature while the air rate adjusted for keeping the regenerator temperature constant which consequences the dependency of both dependent and independent variables on the unit performance. The air flow rate to the regenerator was also an independent variable during the parametric sensitivity analysis and its effect on FCC performance was investigated at constant T_{feed}, F_{feed} and CCR. Combining all the sensitivity analysis, it has been found to increase gas oil conversion and product yields by 5 to 6 percent with decrease of say, 10 K, in the feed preheat temperature (T_{feed}) and corresponding increase in air rate (F_{air}) and catalyst circulation rate (F_{rgc}) at constant reactor outlet temperature (ROT) and regenerated catalyst temperature (T_{rgc}).

INTRODUCTION

Fluid Catalytic Cracking (FCC) is one of the most efficient secondary processes to increase gross refinery margin (GRM) and hence increase profitability as it converts lowpriced heavy feedstock into lighter, more valuable hydrocarbons such as liquefied petroleum gas (LPG) and gasoline at high temperature and moderate pressure in presence of a finely divided silica/alumina based catalyst. One of the important advantages of fluid catalytic cracking is the ability of the catalyst to flow easily between the reactor and regenerator when fluidized with appropriate vapor phases. Due to this fluidization of the catalyst, there is intimate interaction between the catalyst and hydrocarbons leading to more cracking reactions. The conversion and yield pattern strongly depend on the feedstock quality, operating conditions of the riser reactor-regenerator sections and the type of catalyst. These complex interactions coupled with economic importance of the unit have prompted many researchers to put their efforts on the modeling of this unit for better understanding and improved productivity.

The complexity of the gas oil composition which cosists of a very large number of components makes it extremely difficult to characterize these components individually and their kinetics at molecular level. Therefore, the complex reactions occurring in the process are generally described by grouping a large number of compnents known as kinetic lumps and defining the reaction network in terms of these lumps. So far, only a limited number of lumps have been considered by researchers to describe the feed as well as the products. Many kinetic models have been developed using 2 - 6, 8, 10, and 12 lump schemes.

Weekman and Nace [1] developed a first kinetic scheme of catalytic cracking and considered only two lumps; feed and products, which accounted for conversion and gasoline yield in isothermal fixed, moving, and fluid bed reactors. Weekman [2] again developed a model to describe the feed and product yield distribution in terms of three lumped components: the gas oil, the gasoline and the remaining C_4s, dry gas and coke. This model was used to study the effect of reaction time on the products, which showed that the time averaged gasoline yield is always less than the instantaneous gasoline yield because of the smoothing effect of time averaging. Lee et al. [3] proposed a 4 lump kinetic model using coke as a separate lump and estimated the kinetic parameters by using the experimental data from literature [4]. The main advantage of this model is its ability to predict coke production. This four lump model was used for the development of correlations and carry out different parametric studies on various aspects of FCC modeling [5-9]. Five lump kinetic models developed by Ancheyta et al. [10] and Bollas et al. [11] included 7 and 9 rate constants respectively.

The advantage of these models is their ability to predict liquefied petroleum gas (LPG) and dry gas yield separately from the other lumps. Ancheyta et al. [12,13]estimated the reaction parameters in the models by using 3 - 5 lump kinetic scheme and developed some correlations for predicting the effect of feedstock properties on catalytic kinetic parameters. These correlations gave good predictions of gasoline and gas plus coke yields. An integrated reactor-regenerator model, using five kinetic lumps with 9 cracking reactions in the riser reactor has been investigated by Dave and Saraf [14]. A selective deactivation kinetic model is more accurate and realistic than the non-selective model, which has been studied by Corella [15].

In all the 2 - 5 lump models, the feed has been characterized in terms of a single kinetic lump. In more detailed kinetic models, the feed is characterized by several hydrocarbon groups with different lumping schemes. These detailed ten or twelve lump models characterize the feed in terms of 4 different heavy hydrocarbon groups such as paraffins, naphthenes, aromatics and carbon atom substituted aromatic rings [16-19]

Gupta et al. [20,21] developed a new detailed kinetic model based on pseudocomponents and also included a 3-phase, one dimensional heat transfer model for the riser reactor. The kinetic parameters were estimated by a semi-empirical approach based on normal probability distribution. A mathematical model was also developed by Arandes et al. [22] which is useful for predicting the behavior of FCC units both under steady and unsteady conditions. Wei et al. [23] and Wu et al. [24] developed kinetic reactor models using different lumping schemes for both riser and downer in fluid catalytic cracking process. Berry et al. [25] developed a two dimensional hydrodynamic model and coupled it with the four kinetic lump model for the reactor. A four lump kinetic model is also used by Baudrez et al. [26] for decoupled solution method to predict the reactive flow and effect of the reactions on the flow. The proposed method is applied to the steady-state, two-phase gas-solid simulation of a Fluid Catalytic Cracking riser reactor. Zhou et al. [27] developed a kinetic model considering reactant oriented selective deactivation for secondary reactions of FCC gasoline because catalyst deactivation is an important function on FCC performance that affects the secondary reaction on FCC gasoline. This validated model is capable of accurately predicting product distribution for secondary reactions of FCC gasoline over a wide range of operating conditions. Recently, a new approach based on transition state theory has been developed by Lee et al. [28] for kinetic modeling of both thermal and catalytic cracking mechanisms of paraffinic naphtha in the circulating fluidized bed.

A four or five lump kinetic scheme seems satisfactory and reasonable to represent the kinetics of gas oil cracking. One may be tempted to pick a more detailed 10 or 12 lump model but its implementation will require a detailed laboratory analysis of the feed in terms of light and heavy paraffins, naphthenes, aromatics etc., a serious practical limitation indeed.

The effects of sensitive parameters have an important role on the FCC performance. A complete reactor regenerator model is most versatile to understand the sensitivity of individual component on the process, which is not been studied in the existing steady state FCC models. To study the sensitivity analysis of each component on the heat balanced FCC model has been simulated from Dave and Saraf [14] with the modified catalyst deactivation function [16] in the present work.

In this paper, a five lump reaction scheme is used to represent the kinetics of gas oil cracking with two (dense and dilute) phase regenerator model. The objectives are to simulate a continuous reactor-regenerator plant which can be used to predict cracking reaction temperature, feed conversion, product yields, regenerator temperature and amount of deposited coke on the spent catalyst and regenerated catalyst. The product yields are assumed to be a function of reaction temperature, and the coke yield plays an important role in the strong interactions between the riser reactor and the regenerator. Any change in the coke yield in the riser affecting the concentration of coke on spent catalyst, for example due to the change in catalyst circulation rate or a feed composition change, results in changes in the rates of the combustion reactions occurring in the regenerator. These affect regenerator temperature and thus regenerated catalyst temperature. This will again affect the product yields and coke yield in the regenerator, thus indicating the existence of strong interaction between riser reactor and regenerator. The steady state simulation of the FCC unit including both the riser reactor and the regenerator is useful for studying the effect of various operating parameters on the performance of this unit. In the present study riser reactor model is validated by comparing its output with the plant data reported in literature [14]. Regenerator flue gas compositions are also calculated at the exit of both dense and dilute beds. This validated combined model is then used to study the effect of feed preheat temperature on the process by keeping the riser outlet temperature constant. The catalyst circulation rate increases with decreasing the feed preheat temperature to reach the target value of riser outlet temperature. The effect of change in feed flow rate and air flow rate on FCC performance was also examined.

As the FCC is most completed process to understand the effect of each parameter on its performance, the sensitivity analysis may help the refiners to understand the effects of individual parameters to improve the profitability of the process.

REACTOR MODEL

Figure 1 shows a schematic of a modern FCC [29] , in which liquid feed enters at the bottom of riser reactor through feed nozzle system with optimum atomization of feed [7]. The liquid feed droplets come in direct contact with hot catalyst particles from the regenerator and are vaporized. These vapors along with catalyst particles move upwards along the riser height and at the same time vapor feed also starts cracking. The cracking reactions taking place in the riser are endothermic. The catalytic reactions occur in vapor phase. The rate of vaporization of feed in the entry zone of the riser reactor affects the cracking performance of the feed to a great extent. The cracking reaction terminates in the riser reactor because of the deactivation of the catalyst due to coke deposition on the catalyst surface as well as the short contact time between catalyst and vapor hydrocarbons in the riser reactor. The expanding volumes of vapors that are generated are the main driving force to carry the catalyst up the riser. The cracked hydrocarbons are separated from the deactivated catalyst in a separator. However, some thermal and non-selective catalytic reactions continue. A number of refineries are modifying the riser termination devices to minimize these undesirable reactions. Some valuable vapor hydrocarbons adsorbed on the catalyst surface are separated out in a stripper using stripping steam. Cyclones are located at the upper part of the reactor to prevent the catalyst particles to move along with product stream. Finally the cracked hydrocarbons from the reactor are recovered in the main fractionator and gas plant. The main fractionator recovers the heavier products such as light cycle oil and decanted oil, from the gasoline and lighter products. The gas plant separates the main fractionator overhead vapors into gasoline, C_3's, C_4's and fuel gas.

The regenerator is assumed to consist of dense bed and the dilute bed or freeboard region shown inFigure 1. The catalyst activity is recovered in the regenerator dense bed by burning off the coke deposited on the spent catalyst during the cracking reaction. This hot regenerated catalyst is recycled in the riser reactor where it is additionally used as a heat carrier to provide the heat required for endothermic cracking reactions and liquid feed vaporization. The 5-kinetic lump reaction scheme proposed by Bollas et al. [11] and shown in Figure 2 has been adopted in the present study.

There are several assumptions made for the modeling of steady state FCC riser reactor: 1) Gases and catalyst are in plug flow in the riser reactor; 2) Gas oil cracking is a second order reaction but cracking of gasoline and LPG are first order reactions; 3) There are no radial temperature gradients in the gas and solid phases; 4) As the catalyst particles are very small (20 μm - 80 μm) and the vaporized gas oil carries catalyst particles at high velocities, slip factor is assumed to be unity; 5) Dry gases produce no coke; 6) Catalyst deactivation is non-selective and related to coke on catalyst only; and 7) The solid catalyst particles are in thermal equilibrium with the gaseous mixture at all times.

With the above assumptions, the model equations for the riser reactor [30-32] are given in Table 1. Tables 2-5 provide plant operating data, design data and thermodynamic and other data [14,30].

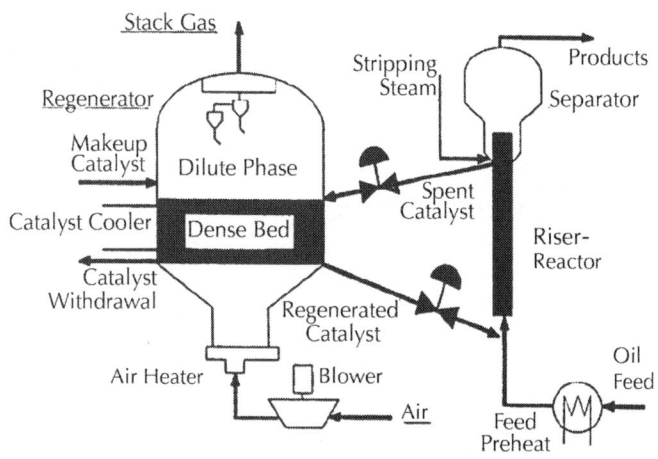

Figure 1: Schematic of modern FCCU (arbel et al. [29]).

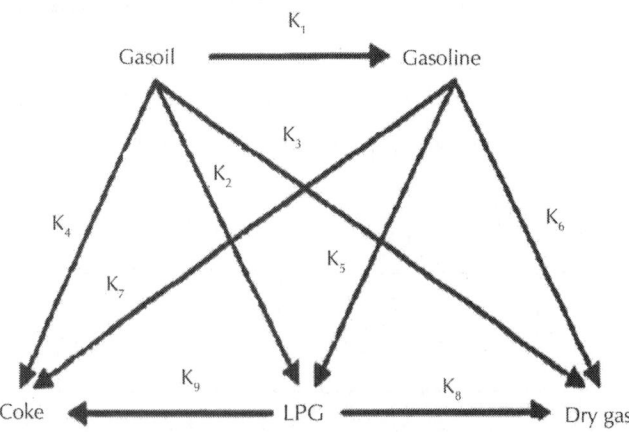

Figure 2: Five lump kinetic scheme.

Table 1: Reactor model equations

The mass balance for the jth lump over a differential element of height dh	
$\dfrac{dF_j}{dh} = A_{ris} H_{ris} (1-\varepsilon) \rho_c \sum_{i=1}^{9} \alpha_{ij} r_i$	(1)
Rate equations for each of the nine reactions is as follows:	
$r_i = k_{0i} \exp\left(-\dfrac{E_i}{RT}\right) C_j^2 \phi \qquad \text{for } i=1,2,3,4 \text{ and } j=1$	(2)
$r_i = k_{0i} \exp\left(-\dfrac{E_i}{RT}\right) C_j \phi \qquad \text{for } i=5,6,7 \text{ and } j=2$	(3)
$r_i = k_{0i} \exp\left(-\dfrac{E_i}{RT}\right) C_j \phi \qquad \text{for } i=8,9 \text{ and } j=3$	(4)
Catalyst deactivation function	
$\phi = \alpha / \left(1 + \beta * t_c^\gamma\right)$ from Jacob $et\ al.$ [16] where $\beta = 12, \gamma = 0.76$	(5)
Enthalpy balance across the same differential element of the riser	
$\dfrac{dT}{dh} = \dfrac{A_{ris} H_{ris} \rho_c (1-\varepsilon)}{F_{rgc} C_{p_c} + F_{feed} C_{p_f v}} \sum_{i=1}^{9} r_i (-\Delta H_i)$	(6)
$T(h=0) = \dfrac{F_{rgc} C_{p_c} (T_{rgn} - 10.0) + F_{feed} C_{p_{fl}} T_{feed} - \Delta H_{evp} F_{feed} - Q_{loss,ris}}{F_{rgc} C_{p_c} + F_{feed} C_{p_{fv}}}$	(7)
Gas oil properties in the riser reactor are calculated by the following equations:	
$MW_g = \sum_{j=1}^{5} X_j MW_j$	(8)
$\rho_v = \dfrac{P_{ris} MW_g}{RT} \quad \text{and} \quad \varepsilon = \dfrac{F_{feed}/\rho_v}{F_{feed}/\rho_v + F_{rgc}/\rho_c}$	(9)
Stripper Model Equations:	
$T_{sc} = ROT - \Delta T_{sc}$	(10)

$$F_{sc} = F_{rgc} + C_{sc} * F_{rgc} \tag{11}$$

Table 2: Input data used in the simulation

Parameters	Numerical Value
F_{feed} (kg/sec)	32.14
F_{rgc} (kg/sec)	208.33
T_{feed} (k)	625.1
P_{ris} (atm)	2.546
P_{rgn} (atm)	2.68
F_{air} (kmol/sec)	0.57
T_{air} (k)	493.9

Table 3: Thermodynamic and other parameters used in the simulation of FCC unit

Parameters	Numerical Value
$C_{p,c}$ (kj/kg·K)	1.003
$C_{p,fl}$ (kj/kg·K)	3.430
$C_{p,v}$ (kj/kg·K)	3.390
$C_{p,N2}$ (kj/kg·K)	30.530
$C_{p,O2}$ (kj/kg·K)	32.280
$C_{p,H2O}$ (kj/kg·K)	36.932
$C_{p,co}$ (kj/kg·K)	30.850
$C_{p,co2}$ (kj/kg·K)	47.400
ΔH_{evp} (kj/kg)	350.0
H_{CO} (kj/kmol)	$1.078 * 10^5$
H_{CO2} (kj/kmol)	$3.933 * 10^5$
H_{H2O} (kj/kmol)	$2.42 * 10^5$
X_{pt}	0.10

ρc (kg/m^3)	1089.0
C_H (kg H$_2$/kg Coke)	0.165
D_p (ft)	$2.0 * 10^{-4}$
MW$_{Gas\ Oil}$	350
MW$_{Gasoline}$	114
MW$_{LPG}$	58
MW$_{Dry\ Gas}$	30
MW$_{Coke}$	12

Table 4: Data used for the simulation of FCC unit

Parameter	Numerical Value
Riser Length (m)	36.965
Riser Diameter (m)	0.684
Regenerator Length (m)	19.344
Regenerator diameter (m)	4.522
Catalyst Inventory in the Regenerator (kg)	34,000
Height of the cyclone inlet (ft)	49

Table 5. Kinetic and thermodynamic parameters used for reactor modeling

Rate Constant	Reaction	Frequency Factor*	Activation Energy (kJ/kmol)	Heat of Reaction (kJ/kmol)
k_1	Gas Oil → Gasoline	18579.9	57,540	45,000
K_2	Gas Oil → LPG	3061.1	52,500	159,315
K_3	Gas Oil → Dry Gas	532.14	49,560	159,315
K_4	Gas Oil → Coke	39.04	31,920	159,315
K_5	Gasoline → LPG	65.4	73,500	42,420
K_6	Gasoline → Dry Gas	0.00	45,360	42,420

K_7	Gasoline → Coke	0.00	66,780	42,420
K_8	LPG → Dry Gas	0.32	39,900	2100
K_9	LPG → Coke	0.19	31,500	2100

REGENERATOR MODEL

The deactivated catalyst, after steam stripping of hydrocarbons, enters the regenerator where all hydrogen in the coke is converted into steam. Carbon can be converted to either CO or CO_2. The heat of combustion raises the temperature of the catalyst recycling from the regenerator. The heat of combustion released in the regenerator is therefore the most critical item in any such simulation.

The following assumptions are made in the development of the regenerator model [14,30,33-35]. 1) The gases are in the plug flow through bed and in thermal equilibrium with surrounding bed; 2) Catalyst in dense bed is well mixed and isothermal with uniform carbon on catalyst; 3) Kinetics of the coke combustion assumes catalyst particles to be 60 μm sizes; 4) Resistance to mass transfer from gas to catalyst phase is negligible; 5) Mean heat capacities of gases and catalyst are assumed to remain constant over the temperature range encountered; and 6) All entrained catalyst is returned via cyclones. The regenerator model equations are given in Table 6.

The CO_2/CO ratio in the gas leaving the dense bed is a function of the bed temperature, residence time, carbonon-catalyst, and equilibrium metals on catalyst. The presence of CO promoters catalyzes CO oxidation and raises CO_2/CO ratio. The CO in the dense bed exit is also oxidized in the dilute bed in presence of entrained catalyst. A set of ordinary differential equations in Table 6(from Equations (23)-(36)) describe the steady state behavior of the gas phase in the regenerator dense bed in terms of the mathematical representation [30].

The Dilute bed is described as a lean phase where entrained catalyst particles and gases evolve in a plug flow pattern. The material and energy balance equations for the dilute bed regenerator are presented (from Equations (45)-(49)) in the Table 6 [30].

Carbon Balance in the Regenerator

The regenerator dense bed consists of two phases, the gases phase and catalyst phase (solid phase), where as the gases are assumed to be moving in plug flow, the catalyst phase is assumed to be well mixed. In this model it is assumed that there is no resistance to mass transfer of gaseous components between gas phase and catalyst phase (Krishna and Perkins [33])

The mathematical representation of carbon balance in the dense bed is given by equation 35 in the Table 6.

STRIPPER MODELING

The aim of stripper is to remove residual hydrocarbons from catalyst surface after cracking reactions. Being a minor unit, no effort was made to rigorously simulate this unit. The spent catalyst temperature and flow rate ware calculated from the model Equations (11) and (12) (shown in the Table 1). A temperature drop of 10 K was assumed across the stripper unit.

SIMULATION PROCEDURE FOR CONTINUOUS REACTOR-REGENERATOR

A continuous riser reactor and regenerator model equations have been coupled by generating a code in C computer language. The ordinary differential equations and nonlinear algebraic equations for material and energy balance (see in Tables 1 and 6) are solved by using a Runge Kutta fourth order and Successive Substitution methods respectively. The calculation of these equations started with initial guess of regenerated catalyst temperature (T_{rgn}) and coke on regenerated catalyst (C_{rgc}), the product yields are calculated at the outlet of the reactor. Subsequently the temperature of spent catalyst and coke on spent catalyst are calculated. The regenerator consists of the two beds: dense bed and dilute bed. The spent catalyst enters into the regenerator dense bed where it regenerates in presence of air and produces flue gases (see Figure 1). The dense bed calculations obtain the new value of catalyst temperature (T_{cal}) and coke on regenerated catalyst (C_{cal}) which is compared with the initial value of T_{rgn} and C_{rgc}. If T_{cal} and C_{cal} do not match with assumed T_{rgn} and C_{rgc} then one needs to start the reactor calculation with newly calculated values of T_{rgn} and C_{rgc} by using the successive substitution method. Finally all the reactor and regenerator equations are solved with converged value of T_{rgn} and C_{rgc}. The tolerance for the convergence of T_{rgn} and C_{rgc} used are $1°C$ and 10^{-4} kg of coke/kg of catalyst respectively.

Table 6: Regenerator model equations

The main combustion reactions in the regenerator are as follows:	
$$C + \frac{1}{2}O_2 \xrightarrow{k_{11}} CO$$	(13a)
$$C + O_2 \xrightarrow{k_{12}} CO_2$$	(13b)
$$CO + \frac{1}{2}O_2 \xrightarrow{k_{13c}} CO_2 \qquad \text{Heterogeneous CO combustion}$$	(13c)
$$CO + \frac{1}{2}O_2 \xrightarrow{k_{13h}} CO_2 \qquad \text{Homogeneous CO combustion}$$	(13d)
$$H_2 + \frac{1}{2}O_2 \xrightarrow{k_{14}} H_2O$$	(13e)
Rate equations for the combustion reactions in the regenerator	

$r_{11} = (1-\varepsilon)\rho_c k_{11} \dfrac{C_{rgc}}{MW_c} P_{O_2} = (1-\varepsilon)\rho_c k_{11} \dfrac{C_{rgc}}{MW_c} \dfrac{f_{O_2}}{f_{tot}} P_{rgn}$	(14)
$r_{12} = (1-\varepsilon)\rho_c k_{12} \dfrac{C_{rgc}}{MW_c} P_{O_2} = (1-\varepsilon)\rho_c k_{12} \dfrac{C_{rgc}}{MW_c} \dfrac{f_{O_2}}{f_{tot}} P_{rgn}$	(15)
$r_{13} = k_{13} P_{O_2} P_{CO} = \left(X_{pt}(1-\varepsilon)\rho_c k_{13c} + \varepsilon k_{13h}\right) P_{O_2} P_{CO} = \left(X_{pt}(1-\varepsilon)\rho_c k_{13c} + \varepsilon k_{13h}\right) \dfrac{f_{O_2} f_{CO}}{f_{tot}} P_{rgn}^2$	(16)
$\left(\dfrac{CO}{CO_2}\right)_{Surface} = \dfrac{k_{11}}{k_{12}} = \beta_c = \beta_{c0}\exp\left(\dfrac{-E_\beta}{RT}\right)$	(17)
$k_c = k_{11} + k_{12} = k_{c0}\exp\left(-\dfrac{E_c}{RT}\right)$	(18)
$k_{11} = \dfrac{\beta_c k_c}{\beta_c + 1} = \dfrac{\beta_c k_{c0}\exp\left(-\dfrac{E_c}{RT}\right)}{\beta_c + 1}$	(19)
$k_{12} = \dfrac{k_c}{\beta_c + 1} = \dfrac{k_{c0}\exp\left(-\dfrac{E_c}{RT}\right)}{\beta_c + 1}$	(20)
$k_{13c} = k_{13c0}\exp\left(-\dfrac{E_{13c}}{RT}\right)$	(21)
$k_{13h} = k_{13h0}\exp\left(-\dfrac{E_{13h}}{RT}\right)$	(22)

1.1. Dense Bed Regenerator:

1) Material Balance:

$\dfrac{df_{O_2}}{dz} = -A_{rgn}\left(\dfrac{r_{11}}{2} + r_{12} + \dfrac{r_{13}}{2}\right)$	(23)
$\dfrac{df_{CO}}{dz} = -A_{rgn}\left(r_{13} - r_{11}\right)$	(24)

$$\frac{df_{co_2}}{dz} = A_{rgn}\left(r_{12} + r_{13}\right)$$	(25)
$$\frac{df_{N_2}}{dz} = 0$$	(26)
Initial Conditions (at z=0) for Dense Bed Modeling:	
$$f_{h_2o} = F_{rgc}\left(C_{sc} - C_{rgc}\right)\frac{C_H}{MW_H}$$	(27)
$$f_{O_2} = 0.21F_{air} - \frac{1}{2}f_{h_2o}$$	(28)
$$f_{co} = f_{co_2} = 0$$	(29)
$$f_{N_2} = 0.79F_{air}$$	(30)
$$f_{tot} = f_{o_2} + f_{co} + f_{co_2} + f_{h_2o} + f_{N_2}$$	(31)
2) Energy Balance:	
$$\frac{dT_{rgn}}{dz} = 0$$	(32)
Heat balance across the regenerator dense bed is given by the following equation:	
$$Q_c + Q_H + Q_{air} + Q_{sc} + Q_{ent} = Q_{rgc} + Q_{sg} + Q_{loss}$$	(33)
Where:	
$$Q_c = f_{co(Zbed)}H_{co} + f_{co_2(Zbed)}H_{co_2}$$	(33a)
$$Q_H = f_{H_2o}H_{H_2o}$$	(33b)
$$Q_{air} = F_{air}C_{P_{air}}\left(T_{air} - T_{base}\right)$$	(33c)

$$Q_{sc} = F_{sc}C_{P_{sc}}\left(T_{sc} - T_{base}\right)$$	(33d)
$$Q_{rgc} = F_{rgc}C_{P_c}\left(T_{rgn} - T_{base}\right)$$	(33e)
$$Q_{sg} = f_{co_2(Zbed)}C_{P_{co_2}} + f_{co(Zbed)}C_{P_{co}} + f_{o_2(Zbed)}C_{P_{o_2}} + f_{H_2O}C_{P_{H_2O}} + f_{N_2} \cdot C_{PN_2}$$	(33f)
$$Q_{ent} = F_{ent}C_{P_c}\left(T_{dil(Zbed)} - T_{base}\right)$$	(33g)
The final equation for the dense bed temperature is	
$$T_{rgn} = T_{base} + \frac{f_{co(Zbed)}H_{co} + f_{co_2}H_{co_2} + f_{H_2O}H_{H_2O} + F_{air}C_{P_{ar}}\left(T_{air} - T_{base}\right) + F_{sc}C_{P_c}\left(T_{sc} - T_{base}\right) + Q_{loss,rgn}}{F_{rgc}C_{P_c} + f_{co_2(Zbed)}C_{P_{co_2}} + f_{co(Zbed)}C_{P_{co}} + f_{o_2}C_{P_{o_2}} + f_{H_2O}C_{P_{H_2O}} + f_{N_2}C_{P_{N_2}}}$$	(34)
$$F_{sc}C_{sc}\left(1 - C_H\right) = F_{rgc}C_{rgc}\left(1 - C_H\right) + \left(f_{CO(Zbed)} + f_{CO_2(Zbed)}\right)MW_c$$	(35)
$$C_{rgc} = \left[F_{sc}C_{sc}\left(1 - C_H\right) - \left(f_{CO(Zbed)} + f_{CO_2(Zbed)}\right)MW_c\right] / \left[F_{rgc}\left(1 - C_H\right)\right]$$	(36)
3) Evaluation of Bed Characteristics:	
$$\rho_g = \frac{P_{rgn}}{RT_{rgn}}$$	(37)
$$u = \frac{F_{air}}{\rho_g A_{rgn}}$$	(38)
$$\varepsilon_{den} = \frac{0.305u_1 + 1}{0.305u_1 + 2} \qquad \text{Ewell and Gadmer [36]}$$	(39)
$$\rho_{c,dense} = \rho_c\left(1 - \varepsilon_{den}\right)$$	(40)
$$\rho_{c,dilute} = \text{Max}\left[0,\left(0.582u_1 - 0.878\right)\right] \text{ (lb/ft3),from McFarlane } et\ al.\ [35]$$	(41)
$$\varepsilon_{dil} = \frac{\rho_{dil}}{\rho_c}$$	(42)

$$F_{ent} = p_{c,dil} A_{rgn} u$$	(43)

4) Dense Bed Height:

The regenerator dense bed height is calculated by the given correlation [35]

$$z_{bed} = \min\left[z_{cyc}, \left(2.85 + 0.8u + \frac{W_{reg} - p_{c,dilute} A_{rgn} z_{cyc}}{A_{rgn} P_{c,dense}} \right) * \left(\frac{1}{1 - pc, dilute/pc, dense} \right) \right]$$	(44)

1.2. Dilute Bed Regenerator:

1) Material Balance

$$\frac{df_{O_2}}{dz} = -A_{rgn}\left(\frac{r_{11}}{2} + r_{12} + \frac{r_{13}}{2} \right)$$	(45)
$$\frac{df_{CO}}{dz} = -A_{rgn}\left(r_{13} - r_{11} \right)$$	(46)
$$\frac{df_{CO_2}}{dz} = A_{rgn}\left(r_{12} + r_{13} \right)$$	(47)
$$\frac{df_{c}}{dz} = -A_{rgn}\left(r_{11} + r_{12} \right)$$	(48)

2) Energy Balance:

$$\frac{dT_{dil}}{dz} = \frac{1}{C_{P_{tot}}}\left(H_{CO}\frac{df_{CO}}{dz} + H_{CO_2}\frac{df_{CO_2}}{dz} \right) = \frac{A_{rgn}}{C_{P_{tot}} f_{tot}}\left[H_{CO}\left(r_{11} - r_{13} \right) + H_{CO_2}\left(r_{12} + r_{13} \right) \right]$$ $$C_{P_{tot}} = \frac{C_{P_{N_2}} f_{N_2} + C_{P_{O_2}} f_{O_2} + C_{P_{CO}} f_{CO} + C_{P_{CO_2}} f_{CO_2} + C_{P_{H_2O}} f_{H_2O} + C_{P_c} F_{ent}}{f_{tot}}$$	(49)

RESULTS AND DISCUSSION

A complete reactor regenerator FCC unit has been simulated using the reactor model equations given by Dave and Saraf [14] and the regenerator model equations from different literature sources

[14,30,33,34,37]. The data on activation energies, frequency factors and heat of reaction (Table 5) are also used from the literature [14]. The plant data and the model predicted data are compared inTable 7 and found to be in a good agreement. This validated model is used for different case studies to check the flexibility of the model.

The reactor model has been coupled with regenerator model and used to study the effect of different independent and dependent parameters on the plant performance. In order to study the effect of changing one independent variable on the reactor performance, all others must be held constant. However, it is important that the reactor operates under steady state condition at all times, and this may require some other variable to be varied simultaneously.

The feed flow rate (F_{feed}) and feed preheat temperature (T_{feed}) are the two key independent variables in the FCC process. The effects of these operating variables on steady state FCC unit performance are calculated by varying air flow rate (F_{air}) and catalyst circulation rate (CCR) to keep either regenerated catalyst temp (T_{rgn}) or ROT constant. The air flow rate to the regenerator was also used as an independent variable and its effect on conversion and yield studied.

Effect of Feed Preheat Temperature on FCC Performance at Constant Feed Flow Rate (F_{feed})

At Constant CCR and Constant Regenerator Temperature (T_{rgn})

Feed preheat temperature plays an important role in controlling the temperature in the riser reactor and hence the cracking reactions (see Figures 3 and 4). Gas oil conversions(32.14 kg/sec) and fixed regenerator temperature (937.5 K).

as well as yield of all the products were found to increase linearly with increase in T_{feed} (Figure 3).Figure 4 shows that reactor outlet temperature (ROT) increases nearly linearly with T_{feed} but air flow rate to the regenerator decreases linearly. This is to be expected in view of the fact that with increased ROT, air flow rate must decrease in order to keep T_{rgn} fixed.

Table 7: Comparison of plant measured and models prediction data

Parameters	Plant Measured	Model Predicted
Reactor Outlet Temp (K)	768.8	769.1
Gas Oil (wt%)	45.6	42.5
Gasoline (wt%)	34.0	36.7
LPG (wt%)	12.4	13.2
DG (wt%)	3.4	3.6
Coke (wt%)	4.6	4.0
Regenerator Temp (K)		937.2
Dense Bed Height (m)	937.5	6.5
Coke on Regenerated Catalyst (wt%)	-	0.12
O_2 (Vol%)	-	1.4
CO (Vol%)	-	10.3
CO_2 (Vol%)	-	6.4
N_2 (Vol%)	-	81.9
Flue Gas Temp (K)	-	939.0
Entrained Cat Flow Rate (kg/sec)	-	13.6

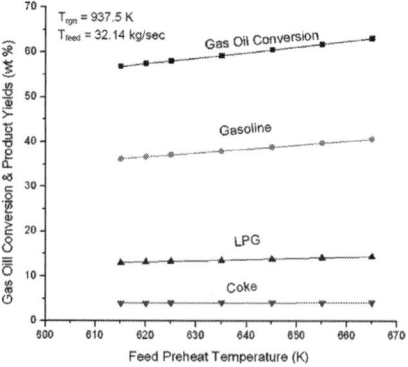

Figure 3: Effect of feed preheat temperature on gas oil conversion and product yields at fixed F_{feed} (32.14 kg/sec) and fixed regenerator temperature (937.5 K).

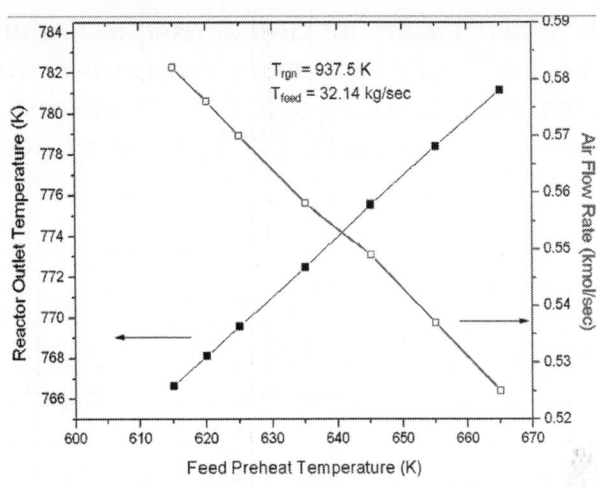

Figure 4: Effect of feed preheat temperature on riser outlet temperature (ROT) at fixed feed

At Constant Air Flow Rate and Constant Reactor Outlet Temperature (ROT)

For ease of operation often the reactor outlet temperature is kept constant with the help of a controller. When feed preheat temperature is increased, regenerated catalyst flow rate (Frgc) must decrease to hold ROT constant (Figure 5). At constant feed rate, this amounts to decreasing cat/oil ratio which leads to decrease in conversion and product yields (Figure 6). Figure 5 also shows that with increasing T_{feed}, regenerator temperature increases initially rapidly and latter gradually. The change in slope seems to occur at feed preheat temperature 625 K perhaps indicating an optimal condition of operation. At low T_{feed} the catalyst circulation rate is high giving rise of high conversion and high rate of coke formation. In view of this, regenerator temperature must increase rapidly, explaining the early sharp rise. From Figure 6 one can conclude that the effect of catalyst circulation rate (or cat/oil) is more pronounced as compared to that of T_{feed}. Increasing T_{feed} alone would have led to increase in conversion. The analysis showed that a decrease in feed preheat temperature by 10 K at fixed ROT and fixed feed could possibly result in 4% increase in gas oil conversion and 3.9 % gasoline yield. This corresponds to an increase in catalyst circulation rate from 208 kg/sec to 220 kg/sec or an increase in cat/oil ratio from 6.5 to 6.9.

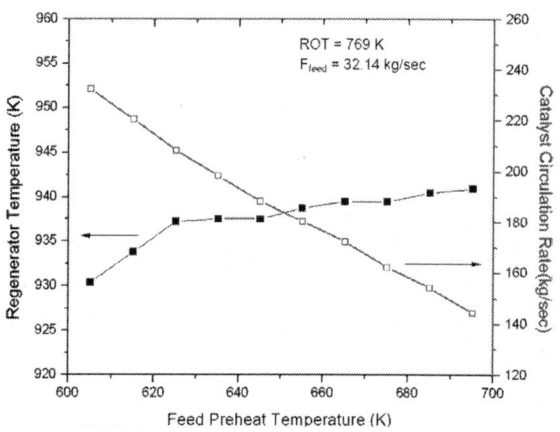

Figure 5: Effect of Feed Preheat Temperature on Regenerator Temperature (T_{rgn}) at Fixed Feed Flow Rate (32.14 kg/sec) and Fix ROT (769 K).

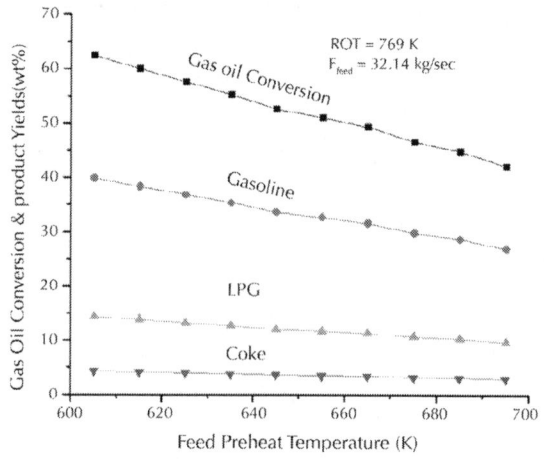

Figure 6: Effect of feed preheat temperature on gas oil conversion and product yields at fixed feed flow rate (32.14 kg/sec) and fix ROT (769 K).

Effect of Feed Flow Rate on FCC Performance at Constant T_{feed}

At Constant CCR and Constant Regenerator Temperature (T_{rgn})

As feed flow rate is increased keeping regenerator temperature and catalyst flow rate constant, the cat/oil ratio decreases which leads to decreased cracking activity and lower conversion and product yields (Figure 7). Figure 8 shows that ROT decreases with increase in feed rate. While lower cat/oil ratio decreases conversion leading to less absorption of endothermic heat, higher feed absorbs more heat. The effect of feed rate being more pronounced as compared to cat/oil ratio, there is net decrease in ROT, which is to be expected since T_{rgn} is fixed. To keep T_{rgn} constant, air flow rate must increase since sensible heat brought in the regenerator by the catalyst is less at lower ROT.

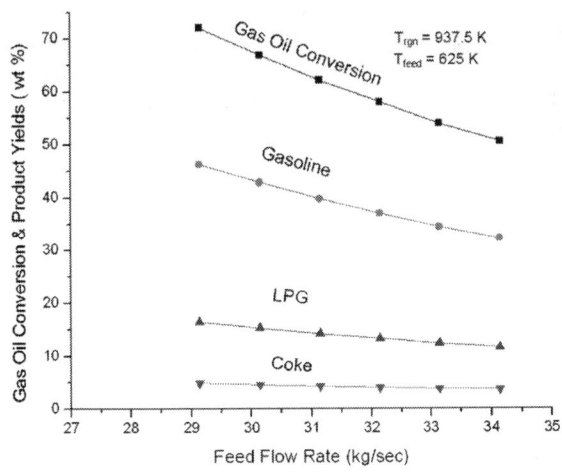

Figure 7: Effect of feed flow rate on the conversion and product yields at fixed feed preheat temperature (625 K) and fixed regenerator temperature (937.5 K).

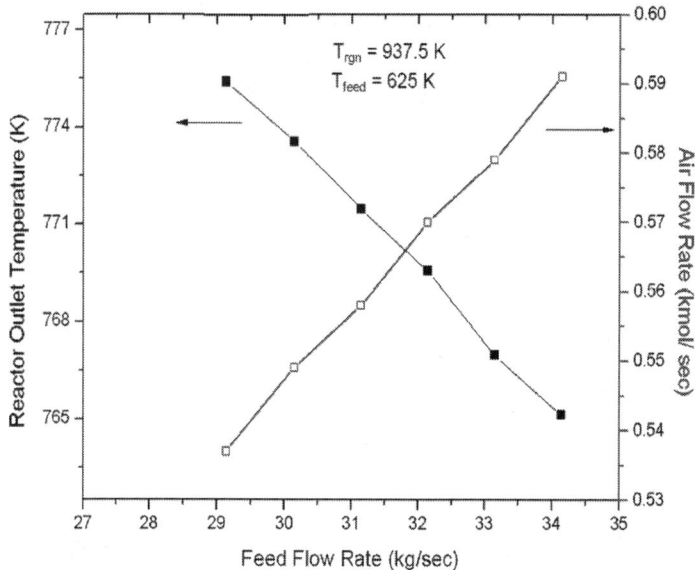

Figure 8: Effect of feed flow rate on the reactor outlet temperature (ROT) at fixed feed preheat temperature (625 K) and fixed regenerator temperature (937.5 K).

At Constant Air Flow Rate and Constant Reactor Outlet Temperature (ROT)

Figure 9 shows effect of change in feed rate on conversion and product yields at constant ROT and air flow rate. Under these conditions, T_{rgn} is expected to decrease because of extra amount of carbon coming in the regenerator (Figure 10). Catalyst circulation rate must increase to keep ROT constant. In the present case both catalyst flow rate and feed rate are increasing, the cat/oil ratio increaseing gradually. This should lead to increase in conversion. However, Figure 9 shows a decreasing trend in conversion as well as product yields. This can be explained in terms of sharp decrease in T_{rgn} amounting to less heat being available for endothermic cracking reactions, particularly when reactor outlet temperature must be maintained constant.

Effect of Air Flow Rate (Fair) on FCC Performance at Constant T_{feed}, F_{feed} and CCR

Figure 11 shows that ROT as well as T_{rgn} increase initially with increasing air rate but become constant at higher air rates. More air rate leads to better coke combustion and hence higher T_{rgn} which in turn, increases ROT. Both T_{rgn} and ROT level off once coke combustion is nearly complete. Higher regenerated catalyst temperature provides higher reactor temperature and hence increased conversion and product yields (Figure 12). These plots suggest that it will be advantageous to increase ROT by 14 K, T_{rgn} by 24 K by increasing air rate to 0.06 kmol/sec.

The result of Figures 11 and 12 have been cross plotted in Figures 13 and 14 which show variation of conversion, product yields and reactor outlet temperature as a function of T_{rgn}.

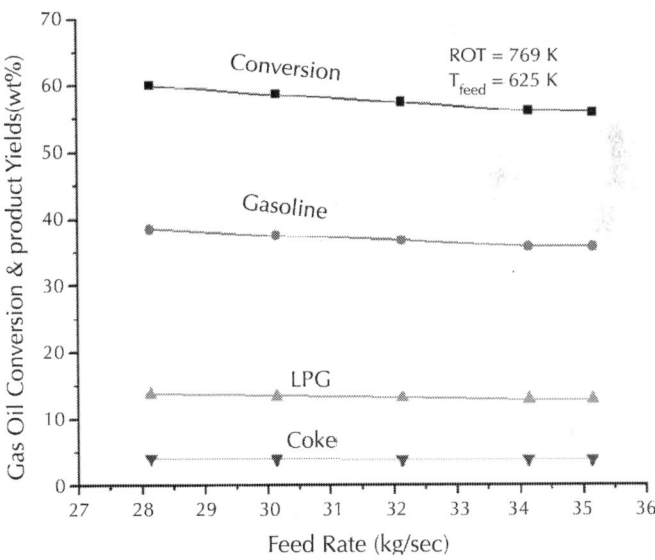

Figure 9: Effect of feed flow rate on the regenerator temperature (T_{rgn}) at fixed feed preheat temperature (625 K) and fixed reactor outlet temperature (768.8 K).

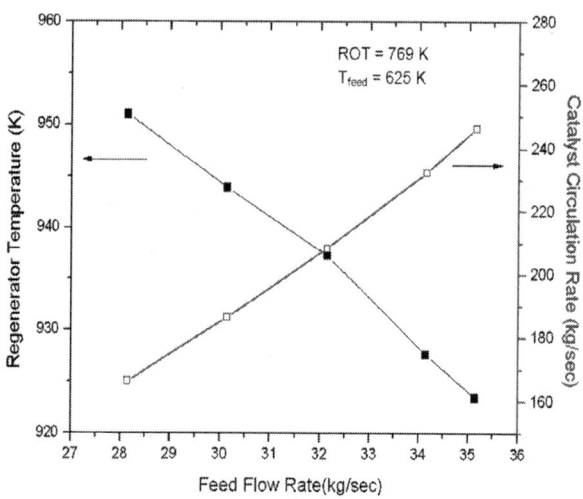

Figure 10: Effect of feed flow rate on the conversion and product yields at fixed reactor outlet temperature (768.8 K) and fixed feed preheat temperature (625 K).

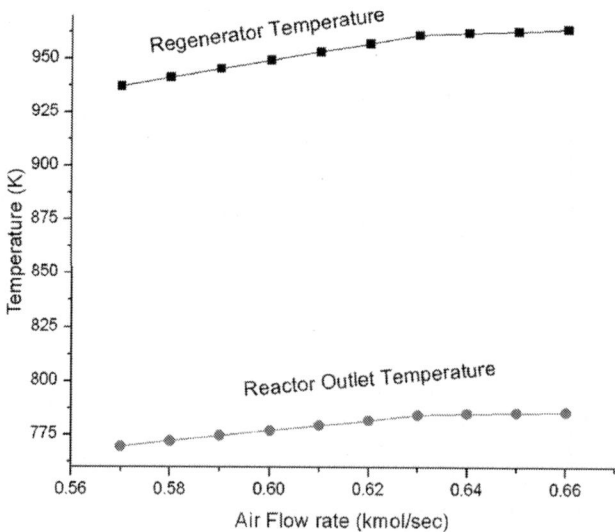

Figure 11: Effect of air flow rate on the regenerator temperature (T_{rgn}) and reactor outlet temperature (ROT).

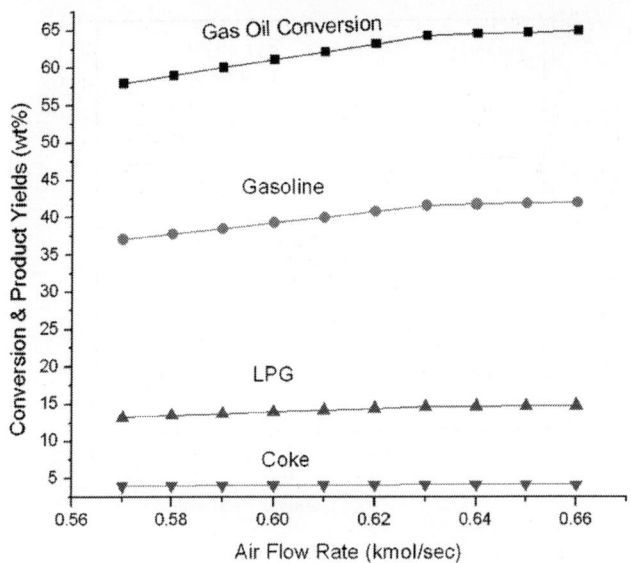

Figure 12: Effect of air flow rate on the conversion and product yields.

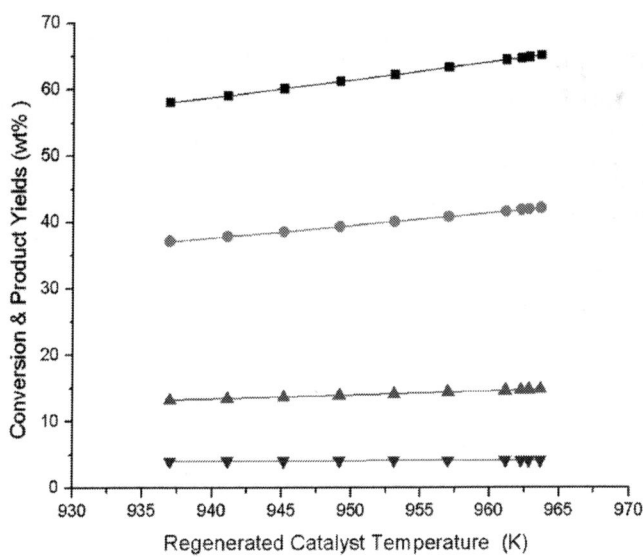

Figure 13: Effect of regenerated catalyst temperature (T_{rgn}) on the conversion and product yields.

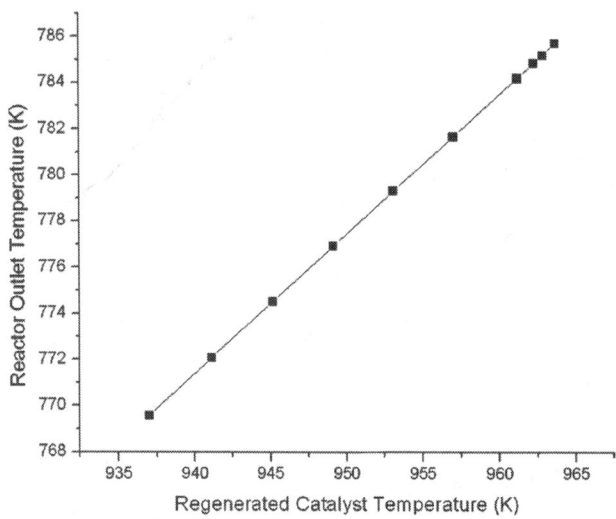

Figure 14: Effect of regenerated catalyst temperature (T_{rgn}) on reactor outlet temperature (ROT).

Combining some of these observations as referred in Table 8, a decrease of say, 10 K, in the feed preheat temperature (T_{feed}) and corresponding increase in air rate (F_{air}) and catalyst circulation rate (F_{rgc}) was found to increase gas oil conversion and product yields by 5 to 6 percent at constant reactor outlet temperature (ROT) and regenerated catalyst temperature (T_{rgc}). The economic visibility of such changes on the operating conditions can be explored by the refiners.

CONCLUSIONS

An industrial FCC unit has been simulated by integrating kinetic models for the riser reactor and the regenerator. The model equations were solved using a computer based code in C-language. The calculated model results are compared with the plant data, which are found to be in agreement. This validated model is used to study parametric sensitivity such as effects of feed preheat temperature, feed flow rate and air flow rate (independent variables) on the FCC performance. Catalyst circulation rate has stronger influence on gas oil conversion as compared to feed preheat temperature for a fixed reactor outlet temperature. On the other hand feed flow rate affects conversion more than catalyst circulation rate. Increase in air flow rate with other important parameters remaining constant leads to increased conversion. From above discussion of sensitivity analysis it appears that decreasing T_{feed} and increasing catalyst circulation rate and air flow rate should lead to higher conversion and product yields. Table 8 shows the result of such computations. At given feed flow rate, a decrease in feed preheat temperature and increase in air flow rate may lead to increased conversion and product yields. However, this will require increased catalyst circulation rate. T_{rgn} and ROT were found to remain essentially constant.

Table 8: Comparison of FCC performance at three different feed preheat temperatures with increased cat/oil ratio and air flow rate

T_{feed} (K)	625.1	615.1	605.1
F_{rgc} (Kg/sec)	208.3	220.3	232.3
F_{air} (Kmol/sec)	0.57	0.63	0.63
ROT(K)	769.1	769.9	771.9
Gas Oil Conversion (wt %)	57.5	60.6	64.2
Gasoline (wt %)	36.8	38.8	41.2
LPG (wt %)	13.2	13.9	14.7
DG (wt%)	3.6	3.8	4.0
Coke (wt %)	4.0	4.2	4.4
T_{rgc} (K)	937.2	936.1	937.2

ACKNOWLEDGEMENTS

The first author thanks the University of Petroleum and Energy Studies for allowing her leave of absence for continuing the research work at University of Saskatchewan, and gratefully acknowledges the financial support of Government of Canada and Indian Oil Corporation Ltd. for this research work.

REFERENCES

1.   V. W. Weekman Jr. and D. M Nace, "A Model of Catalytic Cracking in Fixed, Moving and Fluid Bed Reactors," Industrial and Engineering Chemistry Process Design and Development, Vol. 7, No. 1, 1968, pp. 90-95. doi:10.1021/i260025a018.

2.   V. W. Weekman Jr., "Kinetics and Dynamics of Catalytic Cracking Selectivity in Fixed Bed Reactors," Industrial and Engineering Chemistry Process Design and Development, Vol. 8, No. 3, 1969, pp. 385-391. doi:10.1021/i260031a015.

3.   L. Lee, Y. Chen and T. Huang, "Four-Lump Kinetic Model for Fluid Catalytic Cracking Process," Canadian Journal of Chemical Engineering, Vol. 67, No. 4, 1989, pp. 615-619. doi:10.1002/cjce.5450670414.

4.   I. Wang, "High Temperature Catalytic Cracking," Ph. D. Dissertation, Fuels Engineering Department, University of Utah, Salt Lake City, 1974.

5.   L. C. Yen, R. E. Wrench and A. S. Ong, "Reaction Kinetic Correlation Equation predicts Fluid Catalytic Cracking Coke Yields," Oil & Gas Journal, Vol. 86, 1988, pp. 67-70.

6.   M. A. Abul-Hamayel, "Kinetic Modeling of High—Severity Fluidized Catalytic Cracking," Fuel, Vol. 82, No. 9, 2003, pp. 1113-1118. doi:10.1016/S0016-2361(03)00017-6.

7.   A. Gupta and D. S. Rao, "Model for the Performance of a Fluid Catalytic Cracking (FCC) Riser Reactor: Effect of Feed Atomization," Chemical Engineering Science, Vol. 56, No. 15, 2001, pp. 4499-4503. doi:10.1016/S0009-2509(01)00122-1.

8.   H. Ali, S. Rohani and J. P. Corriou, "Modeling and Control of a Riser-Type Fluid Catalytic Cracking (FCC) Unit," Transactions of the Institution of Chemical Engineers, Vol. 75, 1997, pp. 401-412. doi:10.1205/026387697523868.

9.   A. Blasetti and H. de Lasa, "FCC Riser Unit Operated in the Heat-Transfer Mode: Kinetic Modeling," Industrial & Engineering Chemistry Research, Vol. 36, No. 8, 1997, pp. 3223-3229. doi:10.1021/ie950704v.

10. J. J. Ancheyta, I. F. Lopez and R. E. Aguilar, "5-Lump Kinetic Model for Gas Oil Catalytic Cracking," Applied Catalysis A: General, Vol. 177, No. 2, 1999, pp. 227-235.doi:10.1016/ S0926-860X(98)00262-2.

11. G. M. Bollas, A. A. Lappas, D. K. Iatridis and I. A. Vasalos, "Five-Lump Kinetic Model with Selective Catalyst Deactivation for the Prediction of the Cracking Process," Catalysis Today, Vol. 127, No. 1, 2007, pp. 31-43. doi:10.1016/j. cattod.2007.02.037.

12. J. J. Ancheyta, I. F. Lopez, R. E. Aguilar and M. I. Moreno, "A Strategy for Kinetic Parameter Estimation in the Fluid Catalytic Cracking Process," Industrial & Engineering Chemistry Research, Vol. 36, No. 12, 1997, pp. 5170-5174. doi:10.1021/ie970271r.

13. J. J. Ancheyta, L. F. Lopez and R. L. Aguilar, "Correlations for Predicting the Effect of Feedstock Properties on Catalytic Cracking Kinetic Parameters," Industrial & Engineering Chemistry Research, Vol. 37, No. 12, 1998, pp. 4637-4640.

14. D. J. Dave and D. N. Saraf, "A Model Suitable for Rating and Optimization of Industrial FCC Units," Indian Chemical Engineer, Section (A), Vol. 45, 2003, pp. 7-19.

15. J. Corella, "On the Modeling of the Kinetics of the Selective Deactivation of Catalysts. Application to the Fluidized Catalytic Cracking Process," Industrial & Engineering Chemistry Research, Vol. 43, No. 15, 2004, pp. 4080- 4086. doi:10.1021/ie040033d.

16. S. M. Jacob, B. Gross, S. E. Volts and V. W. Weekman Jr., "A Lumping and Reaction Scheme for Catalytic Cracking," AIChE Journal, Vol. 22, No. 4, 1976, pp. 701-713.doi:10.1002/ aic.690220412.

17. B. Gross, S. M Jacob, D. M. Nace and S. E. Voltz, "Simulation of Catalytic Cracking Process," US Patent 3960707, 1976.

18. S. H. Cerqueira, E. C. Biscaia Jr. and E. F. Sousa-Aguiar, "Mathematical Modeling and Simulation of Catalytic Cracking of Gas Oil in a Fixed Bed: Coke Formation," Applied Catalysis

A: General, Vol. 164, No. 1, 1997, pp. 35-45. doi:10.1016/S0926-860X(97)00155-5.

19. R. C. Ellis, X. Li and J. B. Riggs, "Modeling and Optimization of a Model IV Fluidized Catalytic Cracking Unit," AIChE Journal, Vol. 44, No. 9, 1998, pp. 2068- 2079.doi:10.1002/aic.690440914.

20. R. K. Gupta, V. Kumar and V. K. Srivastava, "Modeling and Simulation of Fluid Catalytic Cracking Unit," Reviews in Chemical Engineering, Vol. 21, 2005, pp. 95- 131. doi:10.1515/REVCE.2005.21.2.95.

21. R. K. Gupta, V. Kumar and V. K. Srivastava, "A New Generic Approach for the Modeling of Fluid Catalytic Cracking (FCC) Riser Reactor," Chemical Engineering Science, Vol. 62, No. 17, 2007, pp. 4510-4528. doi:10.1016/j.ces.2007.05.009.

22. J. M. Arandes, M. J. Azkoiti, J. Bilbao, H. I. de Lasa, "Modeling FCC Units under Steady and Unsteady State Conditions," Canadian Journal of Chemical Engineering, Vol. 78, 2000, pp. 111-123. doi:10.1002/cjce.5450780116.

23. W. Fei, R. Xing, Z. Rujin, L. Guohua and J. Yong, "A Dispersion Model for Fluid Catalytic Cracking Riser and Downer Reactors," Industrial & Engineering Chemistry Research, Vol. 36, No. 12, 1997, pp. 5049-5053. doi:10.1021/ie9702183.

24. C. Wu, Y. Cheng and Y. Jin, "Understanding Riser and Downer Based Fluid Catalytic Cracking Processes by a Comprehensive Two-Dimensional Reactor Model," Industrial & Engineering Chemistry Research, Vol. 48, No. 1, 2009, pp. 12-26. doi:10.1021/ie800168x.

25. T. A. Berry, T. R. McKeen, T. S. Pugsley and A. K. Dalai, "Two Dimensional Reaction Engineering Model of the Riser Section of a Fluid Catalytic cracking Unit," Industrial & Engineering Chemistry Research, Vol. 43, No. 18, 2004, pp. 5571-5581. doi:10.1021/ie0306877.

26. E. Baudrez, G. J. Heynderickx and G. B. Marin, "Steady- state Simulation of Fluid Catalytic Cracking Riser Reactors Using a Decoupled Solution Method with Feedback of the Cracking

Reactions on the Flow," Chemical Engineering Research and Design, Vol. 88, No. 3, 2010, pp. 290-303. doi:10.1016/j.cherd.2009.05.003.

27. X. Zhou, T. Chen, B. Yang, X. Jiang, H. Zhang and L. Wang, "Kinetic Model Considering Reactant Oriented Selective Deactivation for Secondary Reactions of Fluid Catalytic Cracking Gasoline," Energy & Fuels, Vol. 25, No. 6, 2011, pp. 2427-2437. doi:10.1021/ef200316r.

28. J. H. Lee, S. Kang, Y. Kim and S. Park, "New Approach for Kinetic Modeling of Catalytic Cracking of Paraffinic Naphtha," Industrial & Engineering Chemistry Research, Vol. 50, No. 8, 2011, pp. 4264-4279. doi:10.1021/ie1014074.

29. A. Arbel, Z. Haung, I. H. Rinard, R. Shinnar and A. V. Sapre, "Dynamic and Control of Fluidized Catalytic Crackers. 1. Modeling of the Current Generation of FCC's," Industrial & Engineering Chemistry Research, Vol. 34, No. 4, 1995, pp. 1228-1243. doi:10.1021/ie00043a027.

30. R. B. Kasat, D. Kunzuru, D. N. Saraf and S. K. Gupta, "Multiobjective Optimization of Industrial FCC Units Using Elitist Non-Dominated Sorting Genetic Algorithm," Industrial & Engineering Chemistry Research, Vol. 41, No. 19, 2002. pp. 4765-4776. doi:10.1021/ie020087s.

31. Y. X. Sha, "Deactivation by Coke in Residuum Catalytic Cracking, Catalysts Deactivation," In: J. B. Butt, Ed., Bartholomew, Elsevier, Amsterdam, 1991, pp. 327-331.

32. A. Voorhies Jr., "Carbon Formation in Catalytic Cracking," Industrial & Engineering Chemistry, Vol. 37, No. 4, 1945, pp. 318-322.

33. A. S. Krishna and E. S. Parkin, "Modeling the Regenerator in Commercial Fluid Catalytic Cracking Units," Chemical Engineering Progress, Vol. 81, 1985, pp. 57- 62.

34. H. I. De Lasa and J. R. Grace, "The Influence of the Freeboard Region in a Fluidized Bed Catalytic Cracking Regenerator," AIChE Journal, Vol. 25, No. 6, 1979, pp. 984-991. doi:10.1002/aic.690250609.

35. R. C. McFarlane, R. C. Reineman, J. F. Bartere and C. Georgakis, "Dynamic Simulator for a Model IV Fluid Catalytic Cracking Unit," Computers & Chemical Engineering, Vol. 17, 1993, pp. 275-300.

36. R. B. Ewell and G. Gadmer, "Design Cat Crackers by Computer," Hydrocarbon Processing, Vol. 4, 1978, pp. 125-134.

37. A. A. Avidan and R. Shinnar, "Development of Catalytic Cracking Technology. A Lesson in Chemical Reactor Design," Industrial & Engineering Chemistry Research, Vol. 29, No. 6, 1990, pp. 931-942. doi:10.1021/ie00102a001.

Chapter

5

Effect of Preparation Method on Catalytic Properties of Double Perovskite Oxides LaSrFeMo0.9Co0.1O6for Methane Combustion

Jiandong Zheng, Xiongfeng Lang, and
Changjiang Wang

College of Material and Chemical Engineering, Chuzhou University,
Chuzhou, China

ABSTRACT

The double perovskite oxides $LaSrFeMo_{0.9}Co_{0.1}O_6$ was prepared by co-precipitation method and sol-gel method. The title catalysts were calcined at 800°C and characterized by XRD H_2-TPR, SEM and TG-DTA techniques. The catalytic activity was evaluated for methane combustion. The specific surface area of them was calculated by BET model. The samples exhibit significant catalytic activity for methane combustion at 800°C. Upon calcination at 800°C, the LaSrFeMo$_{0.9}$Co$_{0.1}$O$_6$ prepared by sol-gel method retains a specific surface area of 24 m^2·g^{-1} and shows an excellent activity for methane combustion (the conversion of 10% and 90% are obtained at 505°C and 660°C, respectively).

INTRODUCTION

We know nature gas is an important economical energy. Catalytic combustion of nature gas is a crucial technology both for energy production and for environmental pollution abatement [1] - [4]. For heat generation process, using natural gas as fuel, catalytic combustion instead of the conventional combustion has several advantages, such as high efficiency and lower temperature which effectively suppresses thermal NO$_x$ formation. For the effective and stable catalytic combustion process, suitable catalysts play a crucial role. Generally, supported noble metal oxides, particularly palladium oxide, are excellent catalysts for lower temperature combustion, but noble metals are expensive and prone to deactivation owing to sintering, decomposition and undesirable interaction with supports under hydrothermal situations encountered in combustion [5] -[7] . A variety of inexpensive transition metal oxide catalysts, such as solid solution oxides, perovskites, pyrochloresand hexaaluminates have been explored for catalytic combustion of methane [8] - [11].

The perovskite-type oxides have been extensively investigated as catalysts for the combustion of methane with their tailored catalytic property and thermal stability by versatile substitution of Aand/or B-sites. It is well-known that the catalytic properties of perovskites are mainly determined by the nature, oxidation states and relative arrangements of B-site cations. As a subclass of perovskite oxides ABO$_3$, the double perovskites A$_2$B'B''O$_6$ have attracted considerable interest due to their unique structural [12] - [15]. For the double perovskiteoxides, two different B' and B'' cations occupy the B sublattice. They have more variations than the single perovskite-type oxides. The greater variation may promote catalytic chemistry for double perovskite type catalysts. Nevertheless, very few works have been reported in the last decade on exploring the catalytic properties of double perovskites. The double perovskite-oxides A_2FeMoO_6 (A = Ca, Sr and Ba) were studied as catalysts for catalytic combustion of methane in 2004 [16] [17]. Now few papers had been reported the effect of B″ site substitution in the double perovskites $A_2B'B''O_6$ for methane catalytic combustion.

We report here a new double perovskites $LaSrFeMo_{0.9}Co_{0.1}O_6$ prepared by co-precipitation method and solgel method to catalyze combustion of methane. The properties of these catalysts were characterized by XRD, low temperature nitrogen adsorption-desorption (BET), TPR, SEM and TG-DTA techniques. Their catalytic activities were evaluated for methane combustion in a fix bed micro-reactor.

EXPERIMENTAL

Preparation of Materials

Co-Precipitation Method

$LaSrFeMo_{0.9}Co_{0.1}O_6$ double perovskite was prepared by co-precipitation method (carbonates route). The stoichiometric amounts of analytical grade La $(NO_3)_3 \cdot 6H_2O$, Sr $(NO_3)_2$, Fe $(NO_3)_3 \cdot 6H_2O$ and Co $(NO_3)_2 \cdot 6H_2O$ solution were prepared separately by dissolving nitrates in distilled water and MoO_3 in ammonia. The solutions were mixed and added into a well-stirred container by addition of $(NH_4)_2CO_3$ at constant temperature (80°C) and pH value of 7 - 8. After filtering and washing with water several times, the solid product was dried at 120°C for 12 h and then calcined at 800°C in Muffle furnace for 4 h under air. The sample thus prepared is referred to as C1 hereafter.

Sol-Gel Method

The same ingredient $LaSrFeMo_{0.9}Co_{0.1}O_6$ sample was prepared by sol-gel method. The stoichiometric amounts of analytical grade $La (NO_3)_3 \cdot 6H_2O$, $Sr (NO_3)_2$, $Fe (NO_3)_3 \cdot 6H_2O$ and $Co (NO_3)_2 \cdot 6H_2O$ solution were prepared separately by dissolving nitrates in distilled water and MoO_3 in ammonia. The solutions were mixed and added into a well-stirred container by addition of citric acid at constant temperature (60°C) The final solution was evaporated until a gel was formed and then the gel was kept in an oven at 120°C to obtain powder. The resulting powder was calcined at 800°C in Muffle furnace for 4 h under air to form mixed oxide. The sample thus prepared is referred to as S1 hereafter.

Characterization

The phase composition of the calcined samples was determined by X-ray powder diffraction (XRD) using a Ni filter and CuK radiation, at 40 kV and 30 mA. The data were collected between 15° and 70°.

The specific surface areas of the samples were measured on a Gemini V 2380 Series Instrument using N_2 adsorption at liquid N_2 temperature. The specific surface area was determined according to the Brunauer-Emmett-Teller theory.

H_2-TPR measurements were carried out in a flow reactor equipped with a thermal conductivity detector. The samples of 0.05 g were previously pretreated in helium at 400°C for 30 min, and then they were cooled to room temperature. A 5 vol% H_2/N_2 stream (20 ml/min) was passed over the sample while it was heated from 40°C to 800°C at the heating rate of 10°C/min.

The micro-morphologies of testing samples were scrutinized on a field emission scanning electron microscope (SEM, JEOL JSM-6510LV). TG and DSC were carried out on a SDT-Q600 thermal analyzer. Samples were tested over the temperature range from room temperature up to 700°C at the constant heating rate of 10°C/min in air.

Catalytic Activity Test

The reaction of methane combustion was carried out in a conventional microreactor under atmospheric pressure. Catalyst (300 mg, 420 - 841 μm) was loaded in a quartz reactor (i.d. 8 mm), with quartz wool sealed at both ends of the catalyst bed. A mixture of 1 vol% methane and 99 vol% air was fed into the catalyst bed at GHSV = 50,000 h^{-1}. The output gas compositions were analyzed by an on-line gas chromatography (GC9890) with a capillary column and a flame ionization detector (Temperature of column: 150°C, Temperature of sample injector: 200°C, Temperature of detector: 230°C).

RESULTS AND DISCUSSION

Crystalline Phases and Specifics Surface Area

The XRD patterns of two catalysts are shown in Figure 1. It can be seen that the sample have diffraction peaks at 2 = 25°, 30°, 33°, 46°, and 58°, which are typical double perovskite diffraction peaks. The ordered double perovskites $LaSrFeMo_{0.9}Co_{0.1}O_6$ may be regarded as a regular arrangement of alternating FeO_6 and MoO_6/CoO_6 corner-shared octahedra, with La and Sr cations occupying the voids in between the octahedral. A single perovskites of $LaCoO_3$, $LaMoO_3$, $SrMoO_3$ and $SrCoO_3$ can be seen in the Figure 1. It is worthy of notice that those spinel can form double perovskites via solid state reactions. Compared with the sample C1, The catalyst S1 has lower intensity. This implies that the preparation method is important in the formation of double perovskites crystal. Sol-gel method renders the precursor mixture homogenous and enables facile mass transfer, resulting in a formation of pure double perovskites during subsequent calcination at 800°C. The specific surface areas (SSA) of the $LaSrFeMo_{0.9}Co_{0.1}O_6$ samples after calcination at 800°C are shown in Table1 We can see the SSA of S1(24 m^2×g^{-1}) is larger than that of C1(16 m^2×g^{-1}). This it is due to sol-gel method makes the precursor mixture homogenous and enables facile mass transfer.

Temperature-Programmed Reduction by Hydrogen

The H$_2$-TPR profiles of two samples are plotted in Figure 2 to compare their redox properties. It is impossible to reduce the La^{3+} and Sr^{2+} ions at A-site under H$_2$-TPR condition, the H$_2$ consumption peaks displayed on the TPR diagrams of the oxides come from the reduction of B site ions [13]. The curves of LaSrFeMo$_{0.9}$Co$_{0.1}$O$_6$ prepared by sol-gel method have two peaks at 407°C, 575°C, corresponding to Fe^{3+} → Fe^{2+}, MoO$_3$ → MoO$_2$ and Co^{2+} → Co. By contrast, the peak area of C1 is smaller than that of S1. It suggests there has more active oxygen species available for the oxidation reaction of S1. It is in line with the characterization of activity.

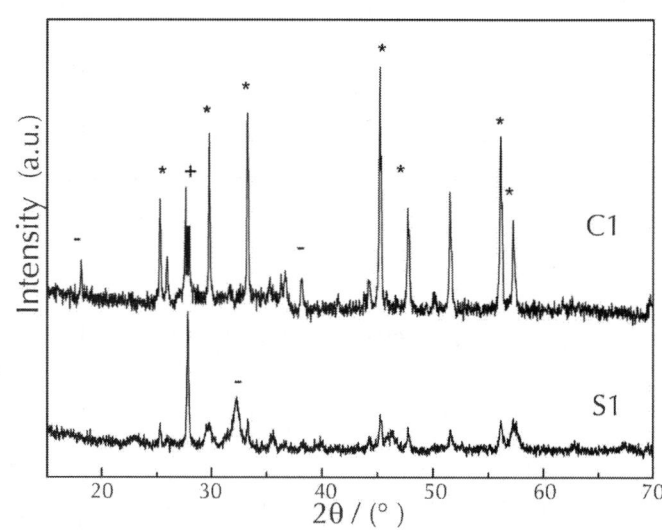

Figure 1: XRD patterns for LaSrFeMo$_{0.9}$Co$_{0.1}$O$_6$ prepared by different methods. *Double perovskite-LaCoO$_3$ SrCoO$_3$ + LaMoO$_3$ SrMoO$_3$.

Table 1: Properties and activities of the samples

Sample	SSA/(m²·g⁻¹)	Catalytic activity in CH$_4$ oxidation/°C		
		T$_{10\%}$	T$_{50\%}$	T$_{90\%}$
C1	16	530	596	684
S1	24	505	575	660

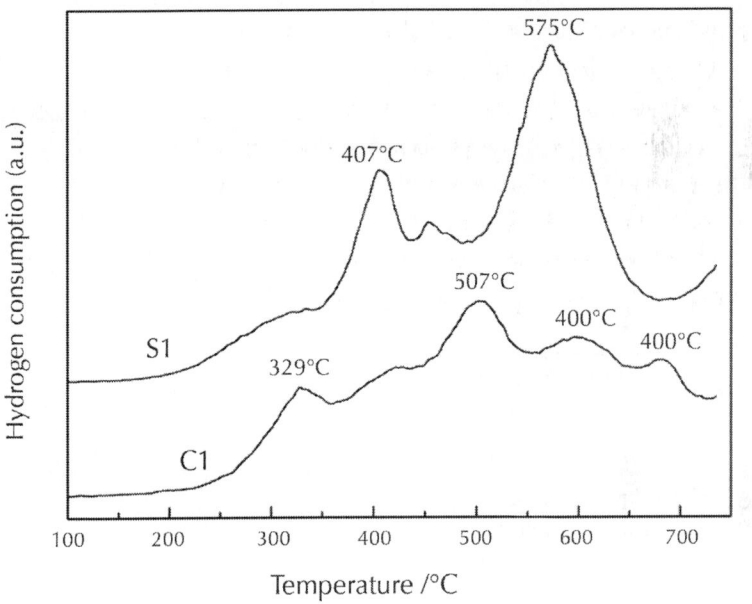

Figure 2: H$_2$-TPR profiles of the LaSrFeMo$_{0.9}$Co$_{0.1}$O$_6$ prepared by different methods.

TG-DSC Analysis

TG-DSC results of C1 and S1 after 120°C ageing are shown in Figure 3. When the samples are heated from room temperature to 800°C at a constant heating rate of 10°C×min⁻¹, C1 and S1 loose about 30% and 70% of weight, respectively. It shows that more organic compounds such as citric acid are removed during the sol-gel process. The samples of C1 and S1 behave differently. For C1 sample, DSC curve presents two lower endothermic peaks and a stronger exothermic peak before 600°C. The first endothermic peak between 250°C and 450°C is due to the removal of the adsorbed water and the decomposition of $SrMnCO_3$, $MnCO_3$ and so on. The loss of weight is above 20% when the temperature is from 100°C to 450°C. The second peak appears when the temperarure is above 550°C. However, the loss of weight is only 6.4%.

By contrast, For S1 sample, DSC curve presents three endothermic peaks before 700°C. The first endothermic peak below 150°C is due to the removal of the excess citric acid and adsorbed water. The second peak between 300°C and 450°C corresponds to the decomposition of carbonate such as $MnCO_3$ and so on. At the same time the loss of weight is 20% when the temperature is from 300°C to 500°C. When the temperature is above 450°C, there is a bigger endothermic peak. However, only 7% weight loss is observed.

So we think, in the process of transformation to form double perovskite oxide, all the losses of weight are due to the removal water, citric acid and the carbonate decomposition. When the temperature is above 550°C, it begins to form double perovskite oxide. In the process of transformation to form double perovskite oxide, it will absorb heat, but no loss arises.

Morphology of Catalysts

Figure 4 presents the SEM images of the catalysts. The catalysts all had the layered structure which can retard the sintering of the material. It was noted that the product of this reaction displayed different particle morphologies with the change in preparation method, from flakes to irregular grains. They were composed of plate-like particles, which were 5 - 10 nm in thickness. From Figure 4, we can see the sample S1 like chrysanthemum. There are lots of honeycombs in the sample S1. It can result that the sample S1 has the higher specific surface areas.

Catalytic Activity

The catalytic behaviors for methane combustion over catalysts are shown in Figure 5 and Table1 All the catalysts, tested under identical experimental conditions, exhibit the S-shaped profiles for CH_4 conversion as a function of reaction temperature. Usually, the activity of methane catalytic combustion was characterized by $T_{10\%}$, $T_{50\%}$ and $T_{90\%}$ representing the reaction temperature at methane conversion of 10%, 50% and 90% respectively. The $T_{10\%}$ and $T_{90\%}$ of C1 catalyst are 530°C and 684°C, respectively. Compared with the C1 catalyst, the S1 catalyst shows higher activity, its $T_{10\%}$ and $T_{90\%}$ are 505°C and 660°C, respectively. Its light-off temperature $T_{10\%}$ decreases by 25°C and total conversion temperature $T_{90\%}$ decreases by 24°C, respectively. It is obvious that the difference in catalytic activity is attributed to the difference in the preparation. Thus we believe that $LaCoO_3$ and $SrCoO_3$ is existed in C1, when the temperature is higher, it will be sintered. Sol-gel method renders the precursor mixture homogenous.

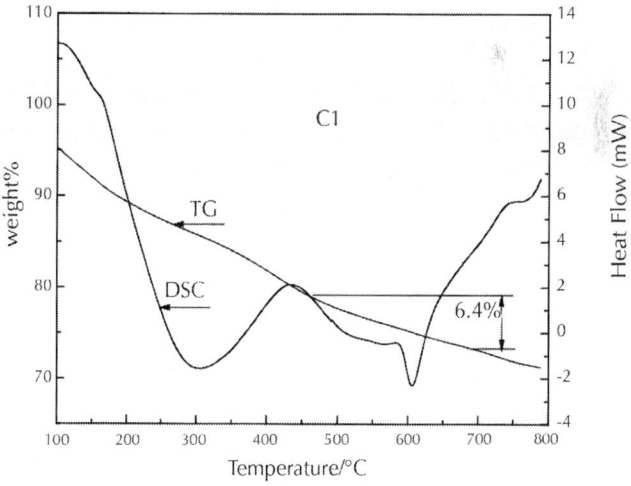

(a)

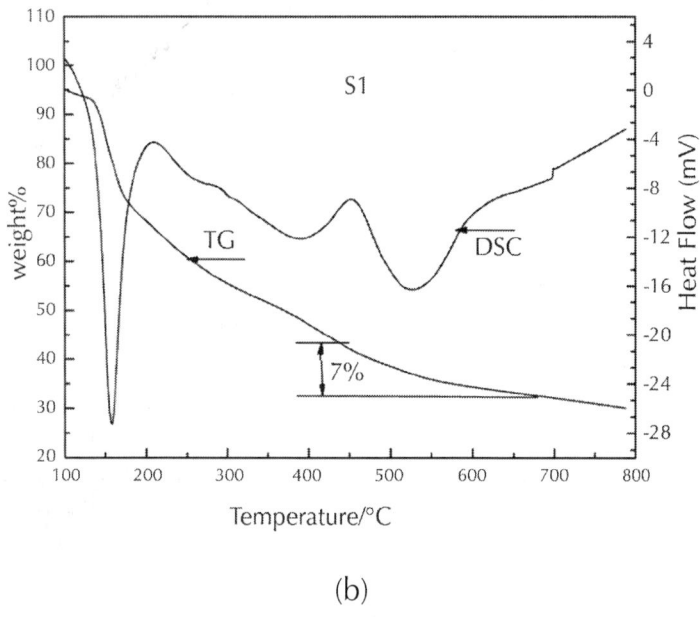

(b)

Figure 3: The TG-DSC curves of LaSrFeMo$_{0.9}$Co$_{0.1}$O$_6$ after 120˚C ageing.

(a)

(b)

Figure 4: SEM images of catalysts after calcination at 800°C.

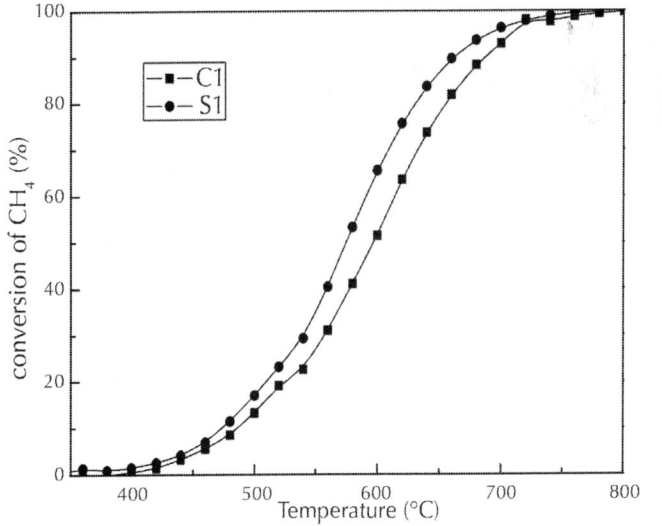

Figure 5: Catalytic activities of catalysts in the combustion of methane.

CONCLUSIONS

The effect of preparation method on the catalytic activity of double perovskite oxides $LaSrFeMo_{0.9}Co_{0.1}O_6$ for methane combustion was investigated. The results showed that $LaSrFeMo_{0.9}Co_{0.1}O_6$ prepared by sol-gel method has better catalytic activity than that prepared by co-precipitation. The catalyst $LaSrFeMo_{0.9}Co_{0.1}O_6$ prepared by sol-gel method possesses the higher specific surface area ($24\ m^2{\cdot}g^{-1}$) and has the lowest temperature for initial and complete conversion of methane than those prepared by co-precipitation method. It could be concluded that preparation method has a great impact on the catalytic activity.

ACKNOWLEDGEMENTS

We are very grateful to the Education Department Natural Science of Anhui Province (KJ2012A213), Project of Teaching Research (2012jyxm534) and the project of Chu Zhou University (2014CXXL054) for the financial support.

REFERENCES

1.   Yan, X. and Crookes, R.J. (2010) Progress in Energy and Combustion Science. Progress in Energy and Combustion Science,   36,   651-676.http://dx.doi.org/10.1016/j.pecs.2010.02.003

2.   Gao, Z. and Wang, R. (2010) Catalytic Activity for Methane Combustion of the Perovskite-Type La1-xSrxCoO3 Oxide Prepared by the Urea Decomposition Method. Applied Catalysis B, 98, 147-153. http://dx.doi.org/10.1016/j.apcatb.2010.05.023

3.   Choudhary, T.V., Banerjee, S. and Choudhary, V.R. (2010) Catalysts for Combustion of Methane and Lower Alkanes.

Applied Catalysis A, 234, 1-23.http://dx.doi.org/10.1016/S0926-860X(02)00231-4

4.   Fabbrini, L. and Rossetti, I. (2003) Effect of Primer on Honeycomb-Supported   La0.9Ce0.1CoO3   Perovskite for Methane Catalytic Flameless Combustion. Applied Catalysis A, 44, 107-116. http://dx.doi.org/10.1016/S0926-3373(03)00025-0

5.   Li, C., Wang, W. and Zhao, N. (2011) Structure Properties and Catalytic Performance in Methane Combustion of Double Perovskites Sr2Mg1-xMnxMoO6. Applied Catalysis B, 102, 78-84. http://dx.doi.org/10.1016/j.apcatb.2010.11.027

6.   Li, C. and Wang, W. (2011) Double Perovskite Oxides Sr2Mg1-xFexMoO6 for Catalytic Oxidation of Methane. Journal of Natural Gas Chemistry, 20, 345-349. http://dx.doi.org/10.1016/S1003-9953(10)60211-3

7.   Li, S. and Wang, X. (2006) Catalytic Combustion of Methane over Mn-Substituted Ba-La-Hexaaluminate Nanoparticles. Journal of Alloys and Compounds, 432, 333-337. http://dx.doi.org/10.1016/j.jallcom.2006.06.022

8.   Todd, H., Gardner, J.J. and Spivey, A.C. (2013) CH4-CO2 Reforming over Ni-Substituted Barium Hexaaluminate Catalysts. Applied Catalysis B, 157, 129-136.

9.   Hu, R.S., Ding, R.R., Chen, J., Hu, J.N. and Zhang, Y.L. (2012) Preparation and Catalytic Activities of the Novel Double Perovskite-Type Oxide La2CuNiO6 for Methane Combustion. Catalysis Communications, 21, 38-41. http://dx.doi.org/10.1016/j.catcom.2012.01.008

10.   Ren, X., Zheng, J. and Song, Y. (2008) Catalytic Properties of Fe and Mn Modified Lanthanum Hexaaluminates for Catalytic Combustion of Methane. Catalysis Communications, 9, 807-810. http://dx.doi.org/10.1016/j.catcom.2007.09.004

11.   Machej, T. and Serwicka, E.M. (2014) Cu/Mn-Based Mixed Oxides Derived from Hydrotalcite-Like Precursors as Catalysts for Methane Combustion. Applied Catalysis A: General, 47, 487-494.

12. Xiong, Y. and Liang, H. (2009) Partial Oxidation of Methane to Syngas over the Catalyst Derived from Double Perovskite (La0.5Sr0.5)2FeNiO6. Applied Catalysis A: General, 371, 153-160. http://dx.doi.org/10.1016/j.apcata.2009.09.044

13. Baylet, A., Royer, S. and Mare, P. (2008) Effect of Pd Precursor Salt on the Activity and Stability of Pd-Doped Hexaaluminate Catalysts for the CH4 Catalytic Combustion. Applied Catalysis B: Environmental, 81, 88-96.

14. Zhang, K., Zhou, G.D. and Li, J. (2009) Effective Additives of A (Ce, Pr) in Modified Hexaaluminate LaxA1-xNiAl11O19 for Carbon Dioxide Reforming of Methane. Catalysis Letters, 130, 246-253. http://dx.doi.org/10.1007/s10562-009-9876-3

15. Wang, Y., Ouyang, J.H. and Liu, Z. (2010) Influence of Dysprosium Oxide Doping on Thermo Physical Properties of LaMgAl11O19 Ceramics. Materials and Design, 31, 3353-3357. http://dx.doi.org/10.1016/j.matdes.2010.01.058

16. Falcon, H. and Barbero, J. (2004) Double Perovskite Oxides A2FeMoO6 (A = Ca, Sr and Ba) as Catalysts for Methane Combustion. Applied Catalysis B, 53, 37-45. http://dx.doi.org/10.1016/j.apcatb.2004.05.004

17. Zheng, J.D. and Ren, X.G. (2008) Catalytic Properties of a (A = Ba, Ca, Sr, and Y) Modified Lanthanum Hexaaluminates for Catalytic Combustion of Methane. Reaction Kinetics and Catalysis Letters, 93, 3-9. http://dx.doi.org/10.1007/s11144-007-5165-6

Reaction in Situ Found In the Synthesis of a Series of Lanthanide Sulfate Complexes and Investigation on Their Structure, Spectra and Catalytic Activity

Zhaoyan Deng[1], Fengying Bai[2], Yongheng Xing[1*],
Na Xing[1], and Liting Xu[1]

[1]College of Chemistry and Chemical Engineering, Liaoning Normal
University, Dalian, China
[2]College of Life Science, Liaoning Normal University, Dalian, China

ABSTRACT

A series of lanthanide sulfates coordination complexes, $Ln_2(SO_4)_3(H_2O)_8$ (**Ln** = Pr (1), Nd (2), Tb (3), Sm (4), Dy (5), Gd (7), Ho (8)), and $EuK(SO_4)_2$ (6), were constructed by the reaction in situ of lanthanide ions (Ln^{3+}) with flexible dodecanedioic acid and rigid aromatic 5-sulfosalicylic acid under hydrothermal conditions. All of them were characterized by elemental analysis, IR spectroscopy, and single-crystal X-ray diffraction. The crystal structures and coordination modes of metal centers and sulfate ions, as well as the novel reaction mechanism and different conditions of lanthanide ions and 5-sulfosalicylic acid to form the series of lanthanide sulfate complexes, were discussed in detail. Solid-state properties for these crystalline materials, such as thermal stability and powder X-ray diffraction have been investigated. Additionally, the photoluminescent characterizations of the complexes 3, 4, 5 and 6, and the catalytic properties of all the complexes about cyclohexane being oxidized into cyclohexanone/cyclohexanol were investigated and compared.

INTRODUCTION

Metal-organic frameworks (MOFs) [1], as a relatively new class of crystalline coordination polymers, have in the past decades become one of the fastest growing fields in chemistry, which is due to the significance both in academia and industry not only for their structural varieties but also for their fascinating potential applications as functional crystalline materials, including gas storage, carbon dioxide capture and renewable catalysts [2-10]. Compared with transition metal, lanthanide-organic frameworks (LMOFs) possess more unique advantages because of the better optics and magnetism characters from the lanthanide themselves and more coordination numbers which is due to the larger radii of the lanthanide atoms [11-16]. In addition, the selection and accommodation of ligands play a key role for the construction of

LMOFs, and therefore, mixed organic ligands, particularly in rigid-flexible ligands, have been proven to be effective and useful to enrich the varieties of lanthanide-organic frameworks because of their specific features: 1) Rigid ligands play important roles in constructing a stable framework and enhance the fluorescence emissions of complexes; 2) The rotation of the flexible ligands increases the variety of configurations of the coordination polymers [17-33].

As we know, a great number of examples about lanthanide-organic framework with rigid aromatic multicarboxylate have been reported, such as benzoic acid [34,35], 1,2-benzene dicarboxylic acid (o-H_2BDC) [36], 1,3-benzene dicarboxylic acid (m-H_2BDC) [37,38], 1,4- benzene dicarboxylic acid (p-H_2BDC) [39,40], benzene-1,3,5-tricarboxylic acid (H_3BTC) [41], and 1,2,4,5-benzenetetracarboxylic acid (H_4BTEC) [42,43]. However, ligands with the interesting functional sulfonic group to construct lanthanide complexes have been rarely explored and only the complexes [Eu(p-Tos) $(H_2O)_7$][p-Tos]$_2$ $(H_2O)_2$ (p-Tos = Toluene-4-sulfonate) [44]; [Ln_2(ad)$_{2.5}$ (BSA)($H_2O)_2$]$_n$(Ln = Sm, Nd; H_2ad = adipic acid; BAS = benzenesulfonate) [45]; [Eu(BSA)(glu)($H_2O)_2$]·H_2O (Ln = Eu, Sm, Ce, Pr, Nd; H_2glu = glutaric acid, HBSA = benzene sulfonic acid); Ln(SSA)($H_2O)_2$ (SSA = 5-sulfosalicylic acid, Ln = Ce, Pr, Nd and Dy) [46]; and [Ln(SSA)($H_2O)_2$]$_n$·nH_2O (SSA = 5-sulfosalicylic acid, Ln = Gd, Sm, Nd, Tb, Eu, Yb and Dy) [47] appear in the present literatures. 5-sulfosalicylic acid, possessing three potential coordinating groups, -COOH, -SO_3H and -OH, can act as the preferential ligand of the lanthanide complexes [48-51]. In addition, although a number of the lanthanide complexes with flexible linkers have been reported, complexes with aliphatic diacid with more than ten carbon atoms have never been reported [52].

Therefore, in our previous work, we selected 5-sulfosalicylic acid and at the same time chose dodecanedioic acid in synthesizing the rigid-flexible lanthanide coordination polymers. Nevertheless, it is not predictable at all to receive the established products, but surprising that there exist two kinds of 2-D or 3-D lanthanide sulfates after the hydrothermal reaction when these two different acids and lanthanide chloride were dissolved in water/ethanol system, and the resultant reaction has never before been reported. Although there are a great number of reports on organic amine template complexes of the lanthanide sulfates, they generally used lanthanide sulfates as the starting material and employed a special organic amine as the structure-directing agent (SDA) or the pH-adjusting agent [53-56]. Hence, one recognizes that organic amine was directly inserted into the structure of the product, and inevitably, as a result, it is hard to produce some single sulfato lanthanide complexes. Moreover, reports on a series of complexes of lanthanide sulfates are very few. Recently, we used a reaction in situ to synthesize a series of the sulfato lanthanide complexes, $Ln_2(SO_4)_3(H_2O)_8$ (Ln=Pr(1), Nd(2), Tb(3), Sm(4), Dy(5), Gd(7), Ho(8)), $EuK(SO_4)_2$ (6) (shown in Scheme 1).

EXPERIMENTAL SECTION

Materials and Methods

All chemicals purchased were of reagent grade or better and were used without further purification. Lanthanide chloride salts were prepared via dissolving 10 g praseodymium oxides with 100 ml 12 M HCl and then evaporating at 100°C until the crystal film formed. The infrared spectra were recorded on a JASCO FT/IR-480 PLUS Fourier Transform spectrometer with pressed KBr pellets in the range 400 - 4000 cm^{-1}. The luminescence spectra were reported on F-7000 FL Spectrophotometer (200 - 800 nm). The elemental analyses were carried out on a Perkin Elmer 240C automatic analyzer. Lanthanide con-tents were analyzed on a Plasma-Spec(I)-AES model ICP spectrometer. X-ray powder diffraction (XRD) data were collected on a Bruker Advance-D8 with Cu-Ka radiation, in the range 5° < 2q < 60°, with a step size of 0.02° (2q) and an acquisition time of 2 s per step.

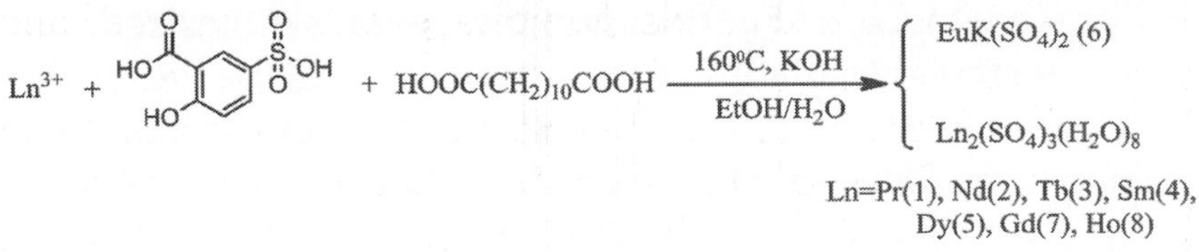

Scheme 1: Synthesis method of complexes 1 - 8.

Synthesis of the Complexes

$[Pr_2(SO_4)_3(H_2O)_8]$ (1). Complexes 1 were prepared by hydrothermal reaction. In a typical synthesis, solution I was prepared by dissolving $PrCl_3 \cdot 6H_2O$ (0.107 g, 0.3 mmol) and 5-sulfosalicylic acid (0.076 g, 0.30 mmol) into 5.0 ml ethanol under stirring for 1 - 2 h, dodecanedioic acid (0.069 g, 0.3 mmol) was added to 5.0 ml ethanol to make solution II. 5.0 ml deionized water was added after solution I was mixed with solution II under stirring for a minimum of 1-2 h. Then, one drop of saturated KOH (aq) was added into the mixture solution under stirring. The final mixture was transferred to a 25 ml Teflon-lined stainless steel vessel under autogenous pressure and heated at 160°C for 3 days. Colorless single crystals of 1 for X-ray diffraction analysis were obtained in ca. 74% yield based on Pr(III) after two weeks. Elemental analysis for $H_{16}O_{20}S_3Pr_2$ ($M_r = 714.16$), calcd: Pr, 39.46; H, 2.26%. Found: Pr, 39.41; H, 2.50%. IR data (KBr pellet, n[cm^{-1}]): 3450.51 (s), 2923.64 (w), 2855.35 (w), 1637.26 (s), 1123.90 (s), 980.17 (w), 651.43 (m), 597.03 (m).

$[Nd_2(SO_4)_3(H_2O)_8]$ (2). This complex was synthesized by a procedure similar to that used for 1 but changing the $PrCl_3 \times 6H_2O$ to $NdCl_3 \times 6H_2O$ (0.108 g, 0.30 mmol), purple crystals of 2 were obtained in ca. 69% yield based on Nd (III). Elemental analysis for $H_{16}O_{20}S_3Nd_2$ ($M_r = 720.82$): Nd: 40.02; H, 2.24%. Found: Nd: 40.00; H, 2.36%. IR data (KBr pellet, n [cm^{-1}]): 3423.01(s), 1642.38 (s), 1130.83 (s), 998.28 (w), 658.03 (m), 599.73 (m).

$[Tb_2(SO_4)_3(H_2O)_8]$ (3). This complex was synthesized by a procedure similar to that used for 1 but changing the $PrCl_3 \times 6H_2O$ to $TbCl_3 \times 6H_2O$ (0.112 g, 0.30 mmol), colorless crystals of 3 were obtained in ca. 78% yield based on Tb(III). Elemental Anal. Calc. for $H_{16}O_{20}S_3Tb_2$ (Mr = 750.20): Tb, 42.37; H, 2.15%. Found: Tb, 42.31; H, 2.42%. IR data (KBr pellet, m[cm^{-1}]): 3487.70 (s), 3371.48 (s), 1642.38 (s), 1143.62 (s), 1092.09 (s), 1001.07 (m), 806.98 (w), 606.12 (s), 489.52 (s), 424.82 (w).

$[Sm_2(SO_4)_3(H_2O)_8]$ (4). This complex was synthesized by a procedure similar to that used for 1 but changing the $PrCl_3 \times 6H_2O$ to $SmCl_3 \times 6H_2O$ (0.109 g, 0.30 mmol), primrose yellow crystals of 4 were obtained in ca. 76% yield based on Sm (III). Elemental analysis for $H_{16}O_{20}S_3Sm_2$ (M_r = 733.06): Sm, 41.02; H, 2.20%. Found: Sm, 39.96; H, 2.51%. IR data (KBr pellet, n [cm^{-1}]): 3448.96 (s), 1642.38 (s), 1124.44 (s), 1007.84 (w), 651.64 (w), 599.73 (s), 489.52 (w).

$[Dy_2(SO_4)_3(H_2O)_8]$ (5). This complex was synthesized by a procedure similar to that used for 1 but changing the $PrCl_3 \times 6H_2O$ to $DyCl_3 \times 6H_2O$ (0.113 g, 0.30 mmol), colorless crystals of 5 were obtained in ca. 71% yield based on Dy (III). Elemental analysis for $H_{16}O_{20}S_3Dy_2$ (M_r = 757.34): Dy, 42.91; H, 2.13%. Found: Dy, 42.83; H, 2.16%. IR data (KBr pellet, n [cm^{-1}]): 3481.31 (s), 3371.48 (s), 3235.32 (m), 1642.38 (s), 1150.02 (w), 1198.49 (s) , 1001.07 (s), 813.37 (m), 748.68 (m), 690.38 (w), 651.64 (w), 606.12 (s), 489.52 (m), 431.22 (m).

$EuK(SO_4)_2$ (6). This complex was synthesized by a procedure similar to that used for 1 but changing the $PrCl_3 \times 6H_2O$ to $EuCl_3 \times 6H_2O$ (0.110 g, 0.30 mmol), primrose yellow crystals of 6 were obtained in ca. 67% yield based on Eu (III). Elemental analysis for $EuK(SO_4)_2$ (M_r = 383.21): Eu, 39.66; K, 10.20%. Found: Eu, 39.61; K, 10.12%. IR data (KBr pellet, n [cm^{-1}]): 3432.24 (w), 2926.42 (w), 2857.74 (w), 1743.66 (w), 1265.24 (w), 1121.91 (s) , 636.34 (s), 602.19 (s), 451.72 (w).

$[Gd_2(SO_4)_3(H_2O)_8]$ (7). This complex was synthesized by a procedure similar to that used for 1 but changing the $PrCl_3 \times 6H_2O$

to $GdCl_3 \times 6H_2O$ (0.111g, 0.30 mmol), colorless crystals of 7 were obtained in ca. 66% yield based on Gd (III). Elemental analysis for $H_{16}O_{20}S_3Gd_2$ (M_r = 746.84): Gd, 42.11%; H, 2.16%. Found: Gd, 42.06; H, 2.25%. IR data (KBr pellet, n [cm^{-1}]): 3481.31 (s), 3371.48 (s), 3235.32 (m), 1648.77 (s), 1143.62 (s), 1092.09 (s), 1001.07 (s), 806.98 (w), 742.28 (w), 651.64 (w), 599.73 (s), 489.52 (m), 431.22 (m).

$[Ho_2(SO_4)_3(H_2O)_8]$ (8). This complex was synthesized by a procedure similar to that used for 1 but changing the $PrCl_3 \times 6H_2O$ to $HoCl_3 \times 6H_2O$ (0.114 g, 0.30 mmol), pink crystals of 8 were obtained in ca. 61% yield based on Ho (III). Elemental analysis for $H_{16}O_{20}S_3Ho_2$ (M_r = 762.20): Ho, 43.28%; H, 2.12%. Found: Ho, 43.15%; H, 2.32%. IR data (KBr pellet, n [cm^{-1}]): 3486.64 (s), 3384.20 (s), 3240.87 (w), 2331.67 (w), 1641.23 (s), 1395.07 (w), 1148.91 (s), 1101.26 (s), 1005.58 (s), 608.94 (m), 479.11 (w), 438.22 (w).

Initial Characterization

Initial characterizations were carried out by elemental analysis, PXRD and IR studies.

PXRD patterns were recorded in the 2θ range 5-50° using Cu-Ka radiation (Bruker Advance-D8), with a step size of 0.02° (2θ) and a count time of 2s per step. As shown in Figures S1-S8, all the peaks presented in the measured patterns closely match the simulated patterns generated from single crystal diffraction data, which confirm the phase purity of the bulk samples.

The IR spectra for the complexes were recorded as KBr pellets. The IR spectra of complexes 1, 2, 3, 4, 5, 7 and 8 exhibit strong and broad absorption bands in the range of 3009 –3679 cm^{-1} and 1640 cm^{-1}, indicating the presence of coordination water moleculars. In 1, 2, 3, 4, 5, 7 and 8, the SO_2 ions adopt μ_3 or μ_2 coordination modes and lead to low site symmetry C_{3v} or C_{2v}. The bands show medium strong intensity at 994 and 607 cm^{-1}. These may be attributable to the symmetric S-O stretching mode (v_1) and the symmetric SO_2

bending mode (v_2). The strong band around 1117 cm^{-1} splitting into two bands, 1142 and 1093 cm^{-1}, may be assigned to the v_3 mode because of the coordination of the free sulfate group to the metals. Compared with that of the above complexes, the IR spectra of complex 6 shows some difference in that there were only two main characteristic bands of 1139 cm^{-1} and 601 cm^{-1}ascribed to the vibration of the sulfate group [57-59].

Single Crystal Structural Determinations

A suitable single crystal of each compound was carefully selected under a cubic microscope and glued to a thin glass fiber for X-ray measurement. Of these reflection data of the complexes, 1-5, 7 and 8 were collected on a Bruker AXS SMART APEX II CCD diffractometer with graphite monochromatized Mo Kα radiation (λ = 0.71073 Å), and that of 6 was collected on a Xcalibur, Atlas, Gemini ultra CCD diffractometer. The data were reduced using SAINTPLUS, and an empirical absorption correction was applied using the SADABS program. The structure was solved and refined using SHELXL-97. All the hydrogen positions were initially located in the different Fourier maps. Final refinement included atomic positions for all the atoms, anisotropic thermal parameters for all the non-hydrogen atoms, and isotropic thermal parameters for all the hydrogen atoms. Details of the structure solution and final refinements for the compounds are given in Tables 1 and 2. The selected bond lengths and bond angles of complexes 3 are listed in Table 3, that of 6 are represented in Table 4, and that of 1, 2, 4, 5, 7 and 8 are shown in the supplement materials

Experiment Set up for Catalytic Oxidation

The oxidation reactions were carried out under air condition (atmospheric pressure) in Schlenk tubes. In a typical experiment, 0.0004 g of the catalysts (complex 1-8) was dissolved in 3.00 ml of desired solvent. Then the required amounts of H_2O_2 (30% H_2O_2 solution) and HNO_3 were added according to this order. Finally,

0.68 g of cyclohexane was added into the solution to make the cyclohexane/catalyst molar ratio equal to 15,000. The reaction solution was stirred for some time at the given temperature.

For the product analysis, 0.03 g of methylbenzene (internal standard) and 1.5 ml of diethyl ether (to extract the substrate and the organic products from the reaction mixture) were added. The obtained mixture was stirred for 10 min and then a sample (0.8 μL) was taken from the organic phase and analyzed by a GC equipped with a capillary column and a flame ionization detector by the internal standard method. Blank experiments confirmed that no cyclohexanol or cyclohexanone were formed in the absence of the metal catalyst under the same conditions.

Table 1: Crystallographic data for complexes 1-4*

Complexes	1	2	3	4
Empirical Formula	$H_{16}O_{20}Pr_2S_3$	$H_{16}O_{20}Nd_2S_3$	$H_{16}O_{20}Tb_2S_3$	$H_{16}O_{20}Sm_2S_3$
M (g·mol^{-1})	714.13	720.79	750.15	733.01
Temperature(K)	273(2)	296(2)	296(2)	296(2)
Crystal system	Monoclinic	Monoclinic	Monoclinic	Monoclinic
Space group	C2/c	C2/c	C2/c	C2/c
a (Å)	13.7004(13)	13.6617(8)	13.5006(17)	13.552(5)
b (Å)	6.8613(7)	6.8366(4)	6.7165(8)	6.757(3)
c (Å)	18.4632(18)	18.4353(12)	18.247(2)	18.272(7)
(deg)	90	90	90	90
(deg)	102.7970(10)	102.6390(10)	102.100(2)	102.320(6)
(deg)	90	90	90	90
V (Å^3)	1692.5(3)	1680.13(18)	1617.8(3)	1634.8(11)
Z	4	4	4	4
D_{calc} (g·cm^{-3})	2.803	2.850	3.080	2.978
Crystal size (mm)	0.17 × 0.14 × 0.09	0.18 × 0.14 × 0.09	0.18 × 0.14 × 0.09	0.18 × 0.14 × 0.09
F(000)	1368	1376	1416	1392
μ (Mo-K)/mm−1	6.158	6.584	9.162	7.599

θ (deg)	2.26 - 24.97	2.26 - 24.99	2.28 - 28.43	2.28 - 25.00
Reflections collected	4108	4034	4886	3914
Independent reflections	1479	1473	1976	1423
R int	0.0231	0.0249	0.0244	0.0231
Parameters	122	123	130	131
(e) (e Å^{-3})	0.633, −0.847	0.889, −0.781	0.776, −1.297	1.012, −0.956
Goodness of fit	1.090	1.125	1.073	1.121
R[a]	0.0207 (0.0228)[b]	0.0209 (0.0219)[b]	0.0253 (0.0311)[b]	0.0180 (0.0186)[b]
wR$_2$[a]	0.0499 (0.0509)[b]	0.0550 (0.0556)[b]	0.0565 (0.0590)[b]	0.0448 (0.0451)[b]

$$^{*a}R = \Sigma \, ||Fo| - |Fc|| / \Sigma |Fo|$$

$wR_2 = [\Sigma(w(Fo^2 - Fc^2)^2 / [\Sigma(w(Fo^2)^2)^{1/2}; [Fo > 4\sigma(Fo)]$ [b]Based on all data.

RESULTS AND DISCUSSION

Synthesis

As shown in Table 5, in the investigation we tried four methods in order to let the lanthanide ions coordinate with 5-sulfosalicylic acid and a flexible ligand, or to coordinate with 5-sulfosalicylic acid and another rigid ligand. The details are as follows. At first, we designed the experimental method (II), for the purpose of the coordination of lanthanide ions and 5-sulfosalicylic acid and chain-like aliphatic dicarboxylic acid to achieve novel complexes with 5-sulfosalicylic acid-dicarboxylic acid as rigid-flexible ligands. But unfortunately, every experimental result demonstrated that lanthanide ions only coordinated with aliphatic dicarboxylic acid. In this case, in order to explore the experimental conditions of the coordination of

5-sulfosalicylic acid and lanthanide carbon chain and large steric hindrance, it maybe not coordinate with lanthanide ions to form stable skeleton of complexes.

Table 2. Crystallographic data for complexes 5-8*

Complexes	5	6	7	8
Empirical Formula	$H_{16}O_{20}Dy_2S_3$	O_8KEuS	$H_{16}O_{20}Gd_2S_3$	$H_{16}O_{20}Ho_2S_3$
M (g·mol^{-1})	757.34	383.20	746.81	762.20
Temperature(K)	296(2)	293(2)	296(2)	296(2)
Crystal system	Monoclinic	Triclinic	Monoclinic	Monoclinic
Space group	C2/c	P1	C2/c	C2/c
a (Å)	13.5034(15)	5.3589(3)	13.574(2)	13.4516(17)
b (Å)	6.7192(7)	6.8831(3)	6.7622(12)	6.6885(8)
c (Å)	18.253(2)	8.9525(11)	18.337(3)	18.171(2)
(deg)	90	97.419(9)	90	90
(deg)	102.049(2)	92.338(7)	102.173(2)	102.000(2)
(deg)	90	91.025(6)	90	90
V (Å^3)	1619.6(3)	327.09(5)	1645.3(5)	1599.1(3)
Z	4	2	4	4
D_{calc} (g·cm^{-3})	3.106	3.891	3.015	3.166
Crystal size (mm)	0.18 × 0.14 × 0.09	0.13 × 0.10 × 0.09	0.18 × 0.14 × 0.09	0.18 × 0.14 × 0.09
F(000)	1424	356	1408	1432
μ (Mo-K)/mm−1	9.646	10.868	8.474	10.320
θ (deg)	3.09 - 24.99	2.99 - 24.99	3.38 - 25.00	2.29 - 28.47
Reflections collected	3792	2042	3814	4543
Independent reflections	1408	1158	1430	1910
R int	0.0392	0.0382	0.0324	0.0458
Parameters	115	109	123	115
(e) (e Å^{-3})	5.885, −6.613	6.535, −1.757	2.758, −2.049	7.564, −6.235
Goodness of fit	1.258	1.052	1.150	1.180

| R^a | 0.0477 (0.0478)[b] | 0.0511 (0.0535)[b] | 0.0384 (0.0397)[b] | 0.0584 (0.0589)[b] |
| $wR_2{}^a$ | 0.1077 (0.1078)[b] | 0.1351 (0.1379)[b] | 0.1101 (0.1113)[b] | 0.1413 (0.1420)[b] |

$$^{*a}R = \Sigma \, ||Fo| - |Fc|| \, / \, \Sigma \, |Fo|$$

$$wR_2 = [\Sigma(w(Fo^2 - Fc^2)^2 \, / \, [\Sigma(w(Fo^2)^2)^{1/2}; [Fo > 4\sigma(Fo)]$$ [b]Based on all data.

Table 3. Selected bond distances (Å) and angles (deg) of complex 3

Bond distances					
Tb-O(5)	2.253(8)	Tb-O(2)	2.316(8)	Tb-O(6)	2.317(9)
Tb-O(4)	2.26(9)	Tb-O(1)	2.366(9)	Tb-O(8)	2.429(8)
Tb-O(3)	2.438(10)	Tb-O(7)	2.487(9)		
Bond angles					
O(5)-Tb-O(2)	143.0(3)	O(5)-Tb-O(6)	80.4(3)	O(2)-Tb-O(6)	126.1(3)
O(5)-Tb-O(4)	88.3(3)	O(2)-Tb-O(4)	79.2(3)	O(6)-Tb-O(4)	70.8(3)
O(5)-Tb-O(1)	147.2(3)	O(2)-Tb-O(1)	69.2(3)	O(6)-Tb-O(1)	79.7(3)
O(4)-Tb-O(1)	109.1(4)	O(5)-Tb-O(8)	99.5(3)	O(2)-Tb-O(8)	75.8(3)
O(6)-Tb-O(8)	141.3(3)	O(4)-Tb-O(8)	147.7(3)	O(1)-Tb-O(8)	80.6(3)
O(5)-Tb-O(3)	69.9(3)	O(2)-Tb-O(3)	73.2(3)	O(6)-Tb-O(3)	134.4(3)

O(4)-Tb-O(3)	74.4(3)	O(1)-Tb-O(3)	140.6(3)	O(8)-Tb-O(3)	78.9(3)
O(5)-Tb-O(7)	73.4(3)	O(2)-Tb-O(7)	133.4(3)	O(6)-Tb-O(7)	74.8(3)
O(4)-Tb-O(7)	143.2(3)	O(1)-Tb-O(7)	76.4(3)	O(8)-Tb-O(7)	68.4(3)
O(3)-Tb-O(7)	125.0(3)				

Table 4: Selected bond distances (Å) and angles (deg) of complex 6*

Bond distances					
Eu-O(4)	2.325(8)	Eu-O(7)#1	2.330(8)	Eu-O(8)#2	2.377(8)
Eu-O(3)	2.438(7)	Eu-O(5)	2.443(9)	Eu-O(1)	2.445(8)
Eu-O(6)#3	2.469(7)	Eu-O(2)	2.566(8)	O(6)-Eu#6	2.469(7)
O(7)-Eu41	2.330(8)	O(8)-Eu#2	2.377(8)		
Bond angles					
O(4)-Eu-O(7)# 1	83.7(3)	O(4)-Eu-O(8)#2	79.3(3)	O(7)41-Eu-O(8)42	82.9(3)
O(4)-Eu-O(3)	121.0(3)	O(7)41-Eu-O(3)	78.6(3)	O(8)#2-Eu-O(3)	150.3(3)
O(4)-Eu-O(5)	70.5(3)	O(7)41-Eu-O(5)	121.9(3)	O(8)#2-Eu-O(5)	137.1(3)
O(3)-Eu-O(5)	72.7(3)	O(4)-Eu-O(1)	136.4(3)	O(7)41-Eu-O(1)	139.5(3)

O(8)42-Eu-O(1)	106.5(3)	O(3)-Eu-O(1)	74.8(3)	O(5)-Eu-O(1)	78.0(3)
O(4)-Eu-O(6)#3	149.2(3)	O(7)#1-Eu-O(6)43	74.0(3)	O(8)#2-Eu-O(6)#3	77.1(3)
O(3)-Eu-O(6)43	75.6(2)	O(5)-Eu-O(6)#3	139.7(3)	O(1)-Eu-O(6)43	70.2(3)
O(4)-Eu-O(2)	85.3(3)	O(7)41-Eu-O(2)	158.8(3)	O(8)#2-Eu-O(2)	77.4(3)
O(3)-Eu-O(2)	122.6(3)	O(5)-Eu-O(2)	70.6(3)	O(1)-Eu-O(2)	55.6(2)
O(6)#3-Eu-O(2)	108.3(3)				

* Symmetry transformations used to generate equivalent atoms: #1 $-x + 2$, $-y$, $-z + 2$ #2 $-x + 1$, $-y$, $-z + 2$ #3 x, y + 1, z #6 x, y$-$1, z.

According to the experimental case above, we can use their insolubility and inertia of coordination with lanthanide ions to promote the coordination of lanthanide ions and 5-sulfosalicylic acid. Based on these points, we chose the mixed solvents of water-ethanol. On one hand, it makes the inert ligand dissolve well in the ethanol and strengthen the possibility of coordination with lanthanide ions.

Table 5: Experimental strategies demonstrating the reaction mechanism

Methods			Materials	Conditions	Products
(I)	$LnC1_3.6H_2O$	H_3SSA		Hydro-thermal (160°C. $H_2OIEtOH$)	Solution (No crystal)

(II)	$LnCl_3 \cdot 6H_2O$	H_3SSA	$HOOC(CH_2)$ $_nCOOH$ $(0 \le n \ge 3)$	Hydro-thermal $(160°C.$ $H_2OlEtOH)$	Ln-dicar-boxylate complexes
(III)	$LnCl_3 \cdot 6H_2O$	H_3SSA	$HOOC(CH_2)$ $_nCOOH$ $(10 \le n \ge 12)$	Hydrother-mal $(160°C$ $H_2OlEtOH)$	Ln-SO_4 complexes
(IV)	$LnCl_3 \cdot 6H_2O$	H_3SSA	3Pz-CQ	Hydro-thermal $(160°C.$ $H_2O/EtOH)$	$Ln(SSA)_3$ $(H_2O)_{1.5}$ complex
H_3SSA:5-sulfo-salicylic acid 3Pz-CQ:2,4,6-tis(3,5-dimeth-yl-1H-pyrazol-1-yl)-1,3,5-triazine					

* H_3SSA: 5-sulfosalicylic acid. 3Pz-CQ: 2,4,6-tis(3,5-dimethyl-1H-pyrazol-1-yl)-1,3,5-triazine

On the other hand, because of the addition of ethanol, it increases the mole concentration of 5-sulfosalicylic acid which easily dissolves in water, and as a result, enhances the molecular collision possibilities and coordination opportunities of 5-sulfosalicylic acid and lanthanide ions. All these promote the lanthanide ions to coordinate with the inert ligand as well as 5-sulfosalicylic acid. But in fact, the results of the experiment of (III) indicated that when a weak acidic ligand was added in the reaction system a decomposition reaction of 5-sulfosalicylic acid occurred, and at the same time, it gave rise to a new reaction in situ, namely, the decomposed product (SO_4^{2-}) coordinated to lanthanide directly. This shows that in the reaction above both dodecanedioic acid and 2,4,6-tis(3,5-dimethyl-1H-pyrazol-1-yl)-1,3,5-triazine played a certain role as a pHadjusting agent and a structure-directing agent for the coordination of 5-sulfosalicylic acid and lanthanide ions. Compared to 2,4,6-tis(3,5-dimethyl-1H-pyrazol-1-yl)-1,3, 5-triazine, it seems that dodecanedioic acid played a more important role for coordination. It might offer some protons for the system to be acidic. Then these protons attached at the juncture of the sulfo-group and phenyl and made the unstable 5-sulfosalicylic to decompose and form the rather stable salicylic acid, and the sulfo-group, which has fallen off from the phenyls, then coordinated with lanthanide ions. The detailed reaction mechanism was speculated in Scheme 2.

Structural Description of Complexes

X-ray diffraction determination results reveal that coordination polymers 1 - 5, 7, 8 are isomorphous, therefore complexes 3 and 6 are taken as the examples to present and describe the structures in detail.

Structure of $Ln_2(SO_4)_3(H_2O)_8$, (Ln = Pr(1), Nd(2), Tb(3), Sm(4), Dy(5), Gd(7), Ho(8))

The structure of 3 reveals that it is a 2-D framework complex. The asymmetric unit of 3 contains one Tb^{3+} atom, one and a half SO_4^{2-} anions and four coordination water molecules. The Tb atom is eight-coordinated to four sulfate ions in the monodentate fashion and four water molecules (Figure 1(a), Figures S11(a)-S16(a)). Of the two framework sulfate ions, one binds to two Tb atoms in a monodentate fashion leaving two terminal S-O bonds, whereas another binds to three metal atoms in a monodentate manner leaving one terminal S–O bond (Scheme 3(d) and (c)). The linkage of the TbO_8 polyhedra and the SO_4 tetrahedra by sharing vertices gives rise to neutral inorganic layers parallel to the bc-plane, containing four-membered and eight-membered rings (Figure 1(b)), (Figures S11(b)-S16(b)).

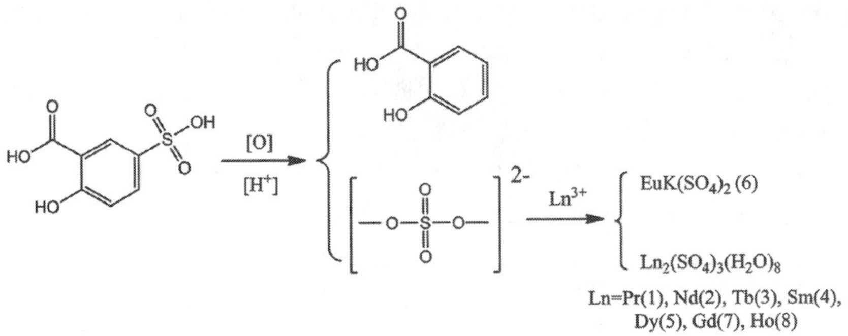

Scheme 2: Reaction mechanism of complexes 1-8.

SO_4^{2-} can be regarded as 2-connected and 3-connected linkers, and Tb^{3+} is surrounded by four SO_4^{2-} ligands and can be regarded as a 4-connected node. Topology analysis by the TOPOS 4.0 software package suggests that complex 3 possesses a 2D (2, 3, 4)-connected 3- nodal network with a Schläfli symbol of $(4^2.6.8^2.10)_2$ $(4^2.6)_2(8)$ (Figure 1(c), Figures S11(c)-S16(c)).

In the complex $Tb_2(SO_4)_3(H_2O)_8$, there exists seven kinds of intermolecular hydrogen bonding interactions of O1-H1A··O9, O1-H1B··O10, O3-H3A··O7, O3-H3B··O8, O4-H4A··O9, O4-H4B··O10, O7-H7B··O8, in which they are respectively formed by water molecular and SO_4^{2-} ligands, except O3-H3A··O7 which is formed by water molecules. In addition, O1-H1A··O9, O4-H4A··O9, O4-H4B··O10 and O7-H7B··O8 linked the adjacent bc planes to generate a 3D framework. At the same timeO1-H1B··O10, O3-H3A··O7, O3-H3B··O8, as the intermolecular hydrogen bondings, further strengthen the whole structure of complexes (Figure 1(d), Figures S11(d)-S16(d)). As shown inTable 6, the hydrogen bonding distances of D···A vary from 2.7151 to 2.9413 Å, and the angles of D-H···A are in the rage of 120.09° - 166.93°. The detail hydrogen bondings of the complexes are listed in Tables S7-S12. All these are comparable with those found in other reported lanthanide sulfate complexes [55,60-62].

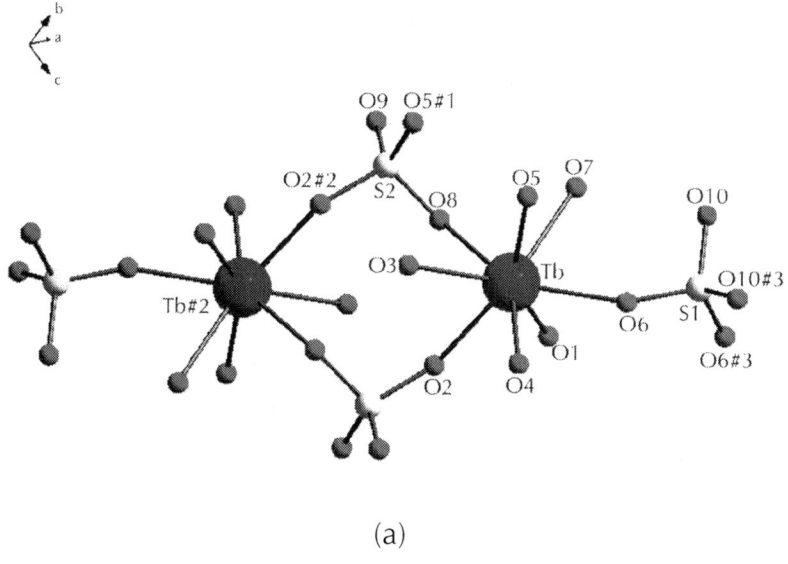

(a)

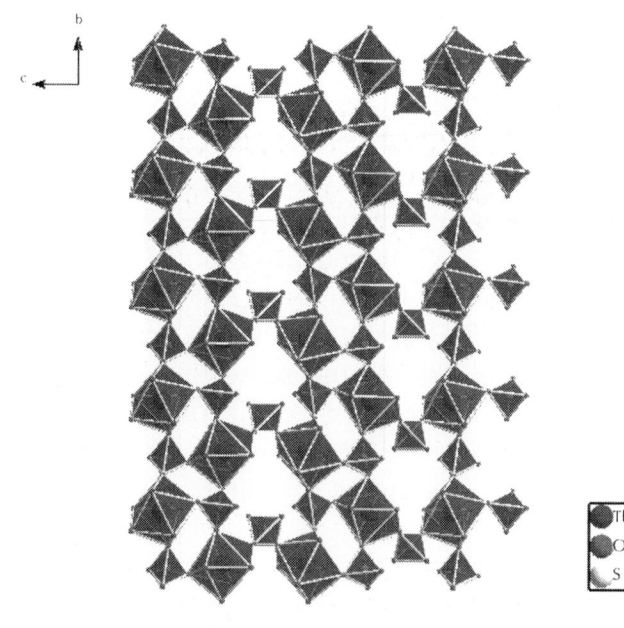

(b)

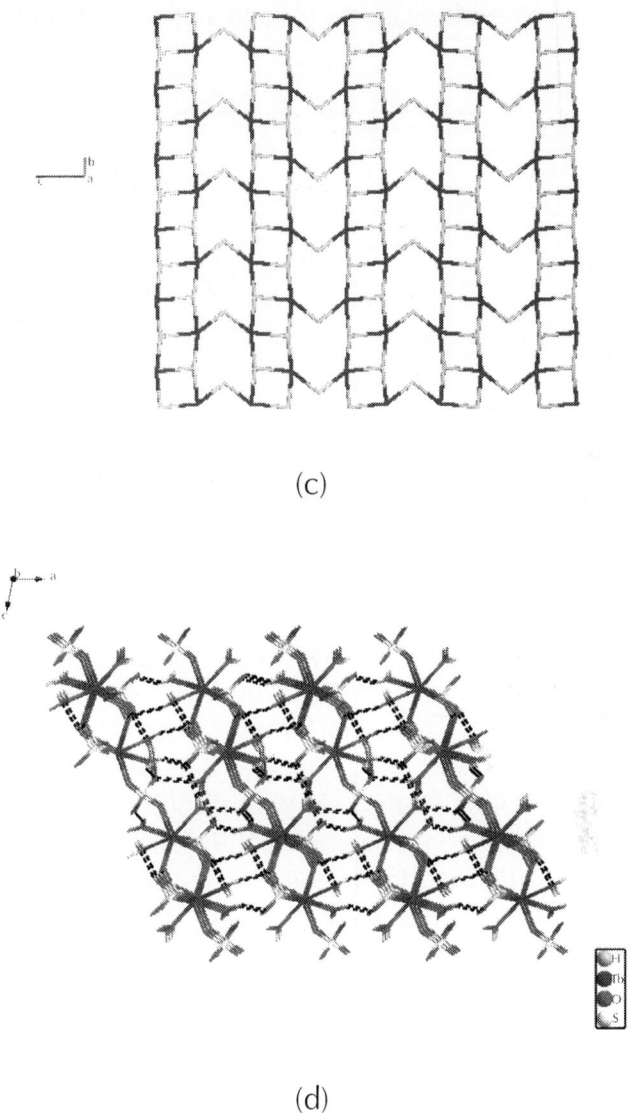

(c)

(d)

Figure 1: (a) The coordination environment of Tb in complex 7; (Symmetry codes: #1: −x+3/2, −y−1/2, −z + 1; #2: −x + 3/2, −y + 1/2, −z + 1; #3: −x + 2, y, −z + 3/2); (b) The 2D layer in complex 3, viewed along the a-axis (c) Schematic representation of the 2D $(4^2.6.8^2.10)_2(4^2.6)_2(8)$ topology network in complex 3. Blue, Tb; Yellow, SO_4^{2-}. (d) A view of the 3D hydrogen bonding network structure of complex 3 (black dotted lines representing hydrogen bonding).

Table 6: Hydrogen bonds distances (Å) and angles (°) of complex 3*

D-H···A	D-H	H···A	D···A	<D-H···A
O1—H1A···O9g	0.9000	1.8777	2.7151	153.93
O1—H1B···O10b	0.9000	1.9754	2.8213	155.95
O3—H3A···O7e	0.7911	2.1652	2.9413	166.93
O1—H3B···O8d	0.7423	2.1431	2.7885	145.81
O4—H4A···O9h	0.8999	2.1300	2.7584	126.16
O4—H4B···O10i	0.9000	2.2402	2.8019	120.09
O7—H7B···O8k	0.9044	2.1543	3.0366	164.89

* Symmetry codes: g: 1/2 + x, −1/2 + y, z; b: x, 1 + y, z; e: 1/2−x, 1/2−y, −z; d: 1/2−x, 3/2−y, −z; h: 1/2−x, −1/2 + y, 1/2−z ; i: −1/2 + x, 1/2 + y,z; k:1−x, 1 + y, 1/2−z.

Structure of $EuK(SO_4)_2$ (6)

The structure of 6 reveals that it is a 3-D framework complex. The asymmetric unit of 6 contains one Eu^{3+} cation, one K^+ cation and two SO_4^{2-} anions. The metal atom is eight-coordinated by the oxygen atoms from seven sulfate ions as can be seen in Figure 2(a). Thus, two of the sulfur atoms in the asymmetric unit S1 forms four S-O-Ln bonds to four crystallographically distinct Ln atoms, thereby sharing corners with four metal-oxygen polyhedra. The S2 forms four S-O-Ln bonds to three unique metal atoms, sharing the corners with two metal-oxygen polyhedra and the edge with another polyhedron (Scheme 3(a) and (b)). The Eu-O bond distances are in the range of 2.325(5) - 2.566(8) Å (av. 2.416(1) Å). The O-Eu-O bond angles are in the range of 55.6(2) - 158.8(3)°. The selected bond distances and angles are given in Table 4. LnO_8 polyhedra are linked by the S(1) O_4tetrahedra in two-dimensions to form fourmembered rings, the rings connected by $S(2)O_4$ to form layers parallel to the bc-plane of the unit cell. Such a connectivity between the four-membered rings results in the formation of six four-membered rings around each eight-membered ring (Figure 2(b)). The layers are stacked over one another in AAA... fashion, with two adjacent layers separated by a unit cell length along the a-axis of the unit cell, thus forming four and eightmembered channels. The $S(1)O_4$ tetrahedra share corners and the $S(2)O_4$ tetrahedra share corners and edges with the metal-oxygen polyhedra respectively from adjacent layers, thereby connecting the layers and forming a threedimensional framework (Figure 2(c)). This connectivity gives rise to a square grid of intersecting four membered channels running perpendicular to one an other, and intersecting the channels along the a-axis of the unit cell. The kalium cations are located in the cages formed by the eight-membered channels.

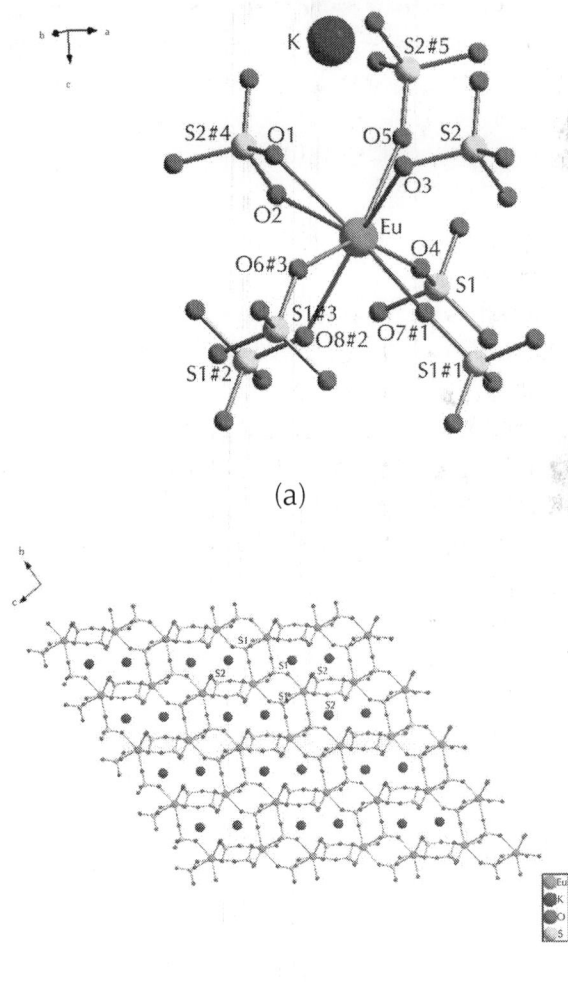

(a)

(b)

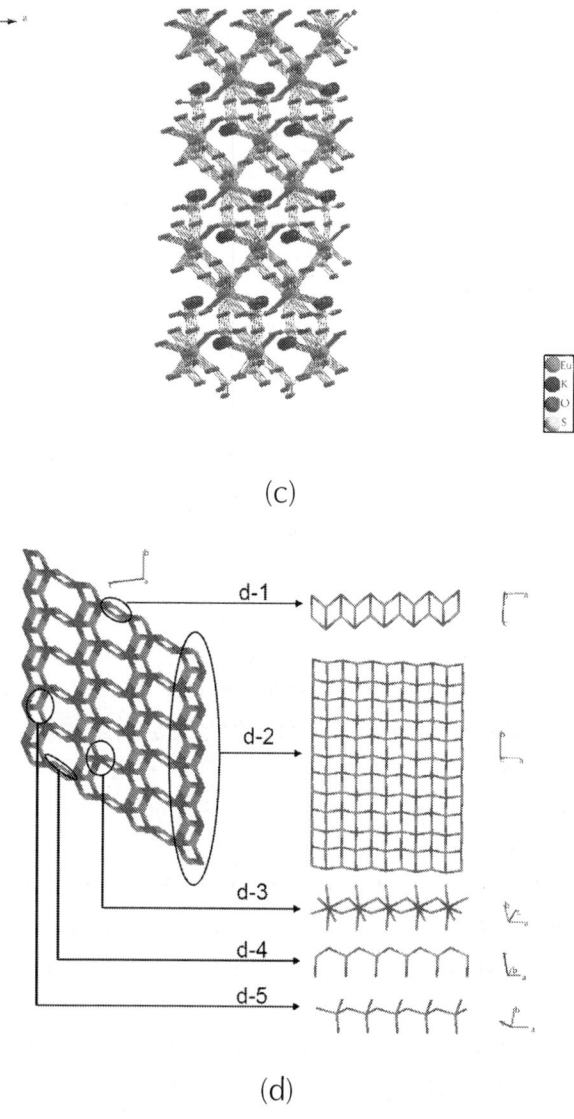

(c)

(d)

Figure 2: (a) The ORTEP view of the coordination environment of Eu in complex 6. (Symmetry codes: #1: −x+2, −y, −z + 2; #2: −x + 1, −y, −z + 2; #3: x, y + 1, z; #4:x−1, y, z; #5:−x + 2,−y, −z + 1); (b) The layer, parallel to the bc-plane of the unit cell in I, formed by connecting EuO_8 polyhedra with SO_4 tetrahedra, by sharing edges and vertices, and thereby forming four-membered and eight-membered rings. Note the arrangement of the four-membered rings around the eight-membered rings. The kalium cations are shown in one eight-membered rings. (c) The three-dimensional framework formed by the linking of the layers, stacked one over another, along the a-axis of the unit cell by the sulfate groups. (d) Schematic representation of the 3D $(4^3)(4^5.6)(4^8.6^9.8^4)$ topology network in complex 6. Fuchia, Eu; Lime, SO_4^{2-}. (d-1) The 1D zig-zig chains in the ac plane. (d-2) The 2D wavelike layer in the ab plane. (d-3) A 7-connected metal mode containing one europium center with seven SO_4^{2-}. (d-4) A 3-connected ligand node containing one SO_4^{2-} with three samarium center. (d-5) A 4-connected ligand node containing one SO_4^{2-} with four europium center.

Each $S(1)O_4$ links four adjacent Eu atoms to afford an infinite wavelike 2D layer in an ab plane (Figure 2(d)-2). Then $S(2)O_4$ between the adjacent layers links three Eu atoms in zig-zig 1D chains in ac planes to construct a 3D framework (Figure 2(d)-1). In this case, $S(1)O_4$ ligands can be regarded as 4-connected nodes (Figure 2(d)-5), $S(2)O_4$ ligands can be regarded as 3-connected nodes (Figure 2(d)-4), and Eu atoms surrounded by seven SO_4^{2-} ligands can be considered as 7-connected nodes (Figure 2(d)-3). Topology analysis by the TOPOS4.0 software package suggests that $EuK(SO_4)_2$ possesses a 3D (3, 4, 7)-connected 3-nodel network with a (4^3) $(4^5.6)$ $(4^8.6^9.8^4)$ Schläfli symbol (Figure 2(d)).

Photoluminescent Properties

Lanthanide coordination polymers often exhibit intense luminescence and thereby are particularly interesting for luminescent materials. Owing to the excellent luminescent properties of Tb (III), Sm (III), Dy (III) and Eu (III) ions, the photoluminescent behaviors of the coordination polymers 3, 4, 5, 6 were investigated in the form of a solid state at room temperature.

The luminescent spectrum of complex 3 was investigated under excitation of 368 nm with a slit width (2.5:2.5). In Figure 3 it can be seen that emission peaks at 491 nm, 545 nm, 587 nm and 622 nm are attributed to the characteristic emissions of Tb emissive state $^5D_4$ to the ground state ^7F_J (J = 6→3), respectively. The spectra are dominated by the $^5D_4 \rightarrow ^7F_5$ transitions at 545 nm which create the intense green luminescence output for the solid sample. In addition, there exists a weak luminescence emission at 417 nm assigned to intraligand $\varpi\,\varpi^*$ transition, implying that the energy is not fully transferred from the sulfate ion to the Tb ion.

The emission spectrum involving complex 4 was determined on excitation at 367 nm with a slit width (2.5:2.5), which is depicted in Figure 4. As expected, the two luminescence emission peaks at 595 and 647 nm correspond to the characteristic emissions of Sm emissive state $^4G_{5/2}$ to the $^4H_{7/2}$ and $^4H_{9/2}$ levels, respectively. Similar to complex 3, there also exists an intraligand $\varpi\,\varpi^*$ transition at 417 nm, implying that the energy is not fully transferred from the sulfate ion to the Sm ion. Compared to those of the other three complexes, the emission spectrum of complex 3 showed very weak and broad.

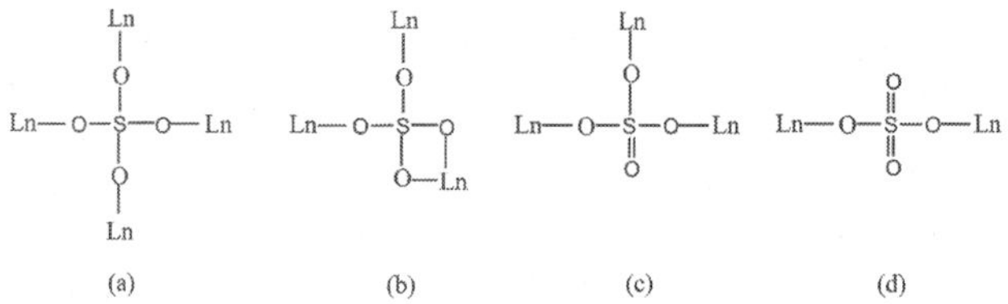

Scheme 3: The four kinds of coordination modes of sulfate ions coordinated to lanthanide ions.

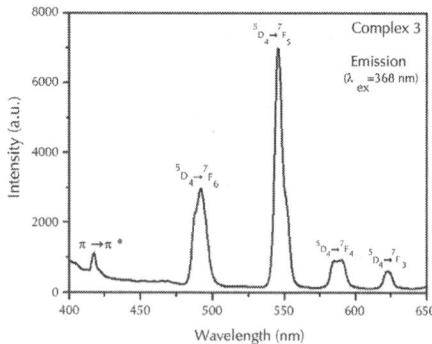

Figure 3: Photoluminescence emission spectra for complex 3 in the solid state at room temperature.

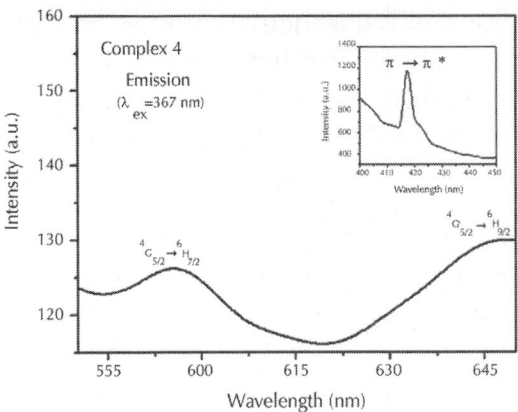

Figure 4: Photoluminescence emission spectra for the complex 4 in the solid state at room temperature.

With regard to the luminescent characteristic of the dysprosium complex, the luminescent spectra of complex 5 was determined under excitation of 240 nm with slit width (2.5:2.5). As shown in Figure 5, it presents good luminescent properties with a narrow, sharp and strong emission peak at 482 nm assigned to the blue insensitive transitions $^4F_{9/2} \rightarrow {}^6H_{15/2}$. Additionally, it should be mentioned that no further emission peaks were observed, indicating that the energy is fully transferred from the sulfate ion to the Dy ion. This is different from the other three complexes in which the characteristic emission bands of the sulfates ions are found, suggesting that the more efficient ligand-to-metal energy transfers occur in dysprosium coordination polymer. Furthermore, the one strong emission band suggests that this homochromy effect possesses good optical application, which is rare in other dysprosium coordination polymers.

The luminescent properties regarding complex 6 were studied at the excitation wavelength of 370 nm with a slit width (2.5:2.5). Figure 6 gives the emission spectra of this complex. The characteristic $^5D_0 \rightarrow {}^7F_J$ (J = 1→4) transitions of the Eu(III) ions at 595, 619, 653, and 689 nm show well an efficient ligand-to-Eu energy transfer. The quite weak emission peak $^5D_0 \rightarrow {}^7F_0$ at 580 nm are attributed to the symmetry-forbidden emission of the Eu(III) ions in these coordination polymers. The $^5D_0 \rightarrow {}^7F_1$ emission bands pertain to the prominent magnetic dipole transitions, which are almost uninfluenced by the coordination environment. On the other hand, the outstanding $^5D_0 \rightarrow {}^7F_2$ emission bands, possessing a strong electric dipole character, are hypersensitive to the coordination environment. Therefore, we can take advantage of the relative intensity disparity of these transitions to probe the nature of the linker environment and herein Eu luminescence can act as a sensitive probe of the lanthanide coordination environment [63,64]. In particular, the ratio of the intensities of the $(^5D_0 \rightarrow {}^7F_2)$: $(^5D_0 \rightarrow {}^7F_1)$ transition is very sensitive to the symmetry of the Eu(III) center. In the spectra, it can be obviously seen that the intensity of the electric dipole transitions $^5D_0 \rightarrow {}^7F_2$ are much stronger than that of the magnetic dipole transition $^5D_0 \rightarrow {}^7F_1$, which implies that the Eu(III) ions in these complexes are located in lower symmetric coordination environments. This is in agreement with the result of the single X-ray analysis. Additionally, the strongest emission peaks in the $^5D_0 \rightarrow {}^7F_1$ and $^5D_0 \rightarrow {}^7F_2$ transition region generate splitting of energy, which can be also ascribed to the Eu(III) centers in asymmetric sites in these complexes. Among these emission lines, $^5D_0 \rightarrow {}^7F_2$ transitions are most striking, indicating intense red luminescence of complex 6. Additionally, there also exits a $\varpi\ \varpi^*$ transition, which is similar to that of the complex 3 and 4.

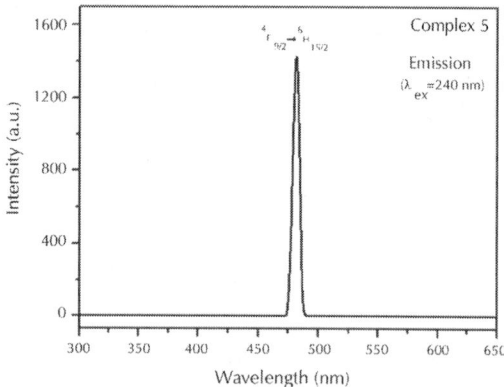

Figure 5: Photoluminescence emission spectrum for complex 5 in the solid state at room temperature.

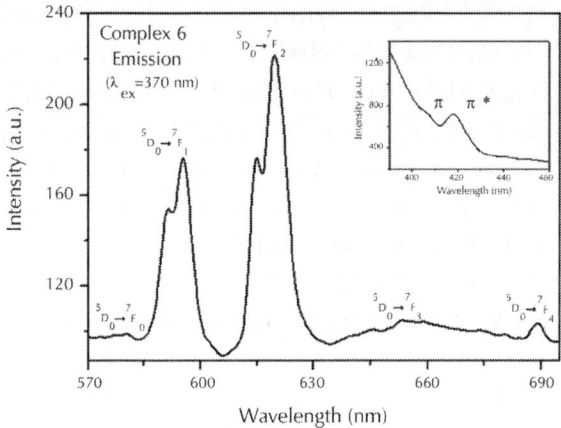

Figure 6: Photoluminescence emission spectra for complex 6 in the solid state at room temperature.

TG Properties

Thermal analysis shows that the complexes are considerably stable, especially complex 6. As shown in Figure 7(a), the weight loss of 20.01% in the range 91°C - 294°C can be attributed to the loss of the coordinated water molecules (cal. 21.07%). Then the complexes stayed at a very stable stage until 800°C. The collapse of the fragment in the second stage may be due to the loss of SO_3, and the final residual is Ln_2O_3. In comparison, the thermal stability of complex 6 is better, and it didn't decompose until 700°C because of the absence of water in the structure (Figure 7(b)).

Catalysis Study

In the primary stage of our work, the oxidation reaction of cyclohexane to produce cyclohexanone and cyclohexanol was employed as a model reaction and complex 1 was examined as a catalyst. The reaction was carried out in a solvent of CH_3CN at 40 centigrade. It showed small TON value, indicating that the capability of the catalysts for the reaction of cyclohexane conversion is weak. A blank experiment conducted in the absence of the catalyst, under the above reaction conditions, found that no products were detected. For the sake of compareson, complexes 2-8 were also used as catalysts. As shown in Table 7, similarly, there is very low catalytic activity with TON values in the range of 0.68 - 4.83, although cyclohexane can reach high conversion after the reaction for some time. It is found that the order of the catalytic activity is 8 > 2 > 5 >3 > 7 > 4 > 6 > 1 when comparing the catalytic activity of the complexes 1-8. In particular, the catalytic efficiency of the complexes 2 and 8 is found to be much better than other complexes. The results also showed that in the cases of 1, 4, 5 and 6, the final products were only cyclohexanol, which indicated that the selectivity of those complexes is very high, and therefore, complexes of 1, 4, 5 and 6 may possess a value for potential application.

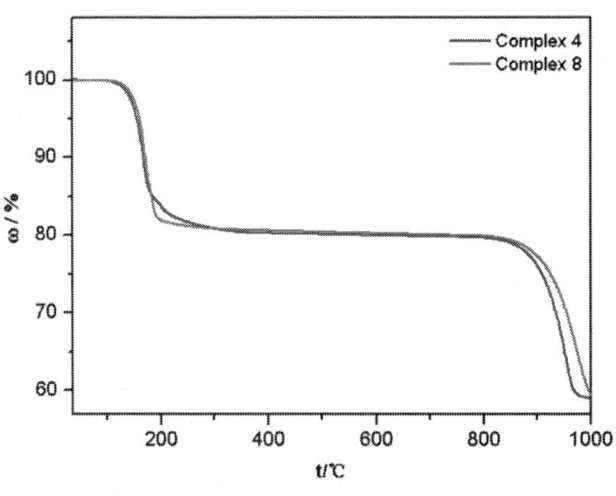

(a)

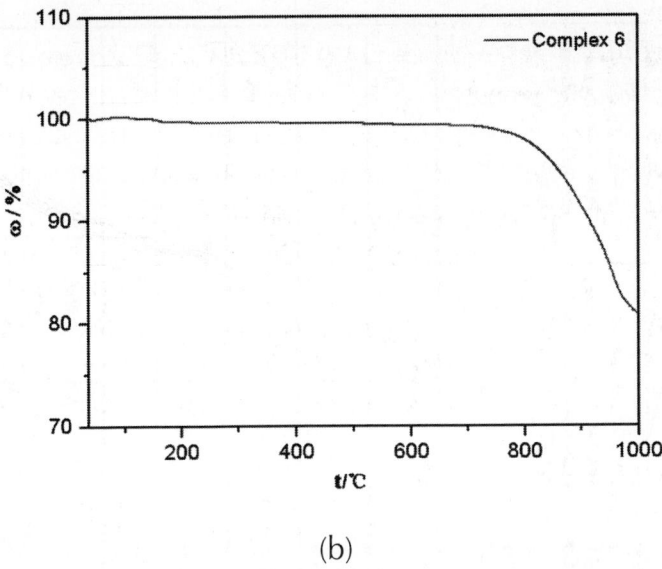

(b)

Figure 7: (a) TG curve of $Ln_2(SO_4)_3(H_2O)_8$ (Ln = Sm(4), Ho(8)); (b) TG curve of $EuK(SO_4)_2(6)$.

Table 7: Data of oxidation for cyclohexane with complexes 1-8 as catalysts in the system of $/H_2O_2/$ HNO_3/CH_3CN at 40°C*

Catalysts	Mole ratio Catalyst/ Cyclohexane (10^{-5})	Cata- lyst/ H_2O_2 $(10^1$	Catalyst $/112 \cdot 10_3$ (10^{-4})	Time (h)	Reaction medium	Cyclohexane Conversion (%)	TON (Cy- clohexa- none)	TON (Cyclo- hexanol)	TON (Cyclohexa- nol and Cyclohexa- none)
1	6.52	1	25	16	CH_3CN	99.88	0	0.68	0.68
2	6.52	1	25	16	CH_3CN	99.95	0.27	4.56	4.83
3	6.52	1	25	16	CH_3CN	99.92	0.06	3.43	3.49
4	6.52	1	25	16	CH_3CN	99.94	0	2.09	2.09
5	6.52	1	25	16	CH_3CN	99.95	0	4.19	4.19
6	6.52	1	25	16	CH_3CN	99.95	0	1.67	1.67
7	6.52	1	25	16	CH_3CN	99.94	0.13	2.51	2.64
8	6.52	1	25	16	CH_3CN	99.94	0.18	4.65	4.83

* 1: $Pr_2(SO_4)_3(H_2O)_8$; 2: $Nd_2(SO_4)_3(H_2O)_8$; 3: $Tb_2(SO_4)_3(H_2O)_8$; 4: $Sm_2(SO_4)_3(H_2O)8$; 5: $Dy_2(SO_4)_3(H_2O)_8$; 6: $Eu(SO_4)_2K$; 7: $Gd_2(SO_4)_3(H_2O)_8$; 8: $Ho_2(SO_4)_3 (H_2O)_8$. TON: moles of products (cyclohexanol and cyclohexanone)/moles of catalysts.

CONCLUSIONS

In summary, we have reported two kinds of simple, high-yield and pure phase synthesis of 2-D and 3-D framework lanthanide sulphate coordination complexes, $Ln_2(SO_4)_3(H_2O)_8$ (Ln = Pr (1), Nd (2), Tb (3), Sm (4), Dy (5), Gd (7), Ho (8)) and $EuK(SO_4)_2$ (6). The synthesis of all complexes have demonstrated that the lanthanide sulfate complexes were designed and prepared by using appropriate solvent and SDAs. On the other hand, all these complexes represent a certain activity in catalysis. These Tb, Sm, Dy and Eu complexes exhibit different luminescence characterizations of Tb^{3+}, Sm^{3+}, Dy^{3+} and Eu^{3+}, respectively. The luminescence in the solid state indicates the complex 5 is an excellent candidate for pure fluorescent materials.

ACKNOWLEDGEMENTS

This work was supported by grants from the National Natural Science Foundation of China (Grant No. 21071071) and the State Key Laboratory of Inorganic Synthesis and Preparative Chemistry, College of Chemistry, Jilin University, Changchun 130012, P. R. China (Grant No. 2013-05).

REFERENCES

1.   Zhou, H.C., Long, J.R. and Yaghi, O.M. (2012) Introduction to metal-organic frameworks. Chemical Reviews, 112, 673-674. http://dx.doi.org/10.1021/cr300014x

2.   Rosi, N.L., Ecket, J., Eddaoudi, M., Vodak, D.T., Kim, J., O'Keeffe, M. and Yaghi, O.M. (2003) Microporous metal-organic frameworks. Science, 300, 1127-1129.http://dx.doi.org/10.1126/science.1083440

3.   Dincä, M., Dailly, A., Liu, Y., Brown, C.M., Neumann, D.A.and Long, J.R. (2006) Hydrogen storage in a microporous metal-organic framework with exposed Mn^{2+} coordination sites. Journal of the American Chemical Society, 128, 16876-16883.http://dx.doi.org/10.1021/ja0656853

4.   Collins, D.J. and Zhou, H.C. (2007) Hydrogen storage in metal-organic frameworks. Journal of Materials Chemistry, 17, 3154-3160. http://dx.doi.org/10.1039/b702858j

5.   Lin, J.D., Cheng, J.W. and Du, S.W. (2008) Five d^{10} 3D metal-organic frameworks constructed from aromatic polycarboxylate acids and flexible imidazole-based ligands. Crystal Growth and Design, 8, 3345-3353. http://dx.doi.org/10.1021/cg8002614

6.   Sun, Y.Q., Zhang, J. and Yang, G.Y. (2006) A series of luminescent lanthanide-cadmium-organic frameworks with helical channels and tubes. Chemical Communications, 4700-4702. http://dx.doi.org/10.1039/b610892j

7.   Kuppler, R.J., Timmons, D.J., Fang, Q.R., Li, J.R., Makal, T.A., Young, M.D., Yuan, D.Q., Zhao, D., Zhuang, W.J. and Zhou, H.C. (2009) Potential applications of metalorganic frameworks. Coordination Chemistry Reviews, 253, 3042-3066.http://dx.doi.org/10.1016/j.ccr.2009.05.019

8.   Li, J.R., Kuppler, R.J. and Zhou, H.C. (2009) Selective gas adsorption and separation in metal-organic frameworks. Chemical Society Reviews, 38, 1477-1504.http://dx.doi.org/10.1039/b802426j

9.   Férey, G. (2008) Hybrid porous solids: Past, present, future. Chemical Society Reviews, 37, 191-214. http://dx.doi.org/10.1039/b618320b

10.   Furukawa, H., Ko, N., Go, Y.B., Aratani, N., Choi, S.B., Choi, E., Yazaydin, A.O., Snurr, R.Q., O'Keeffe, M., Kim,

J. and Yaghi, O.M. (2010) Ultra-high porosity in metal-organic frameworks. Science, 239, 424-428. http://dx.doi.org/10.1126/science.1192160

11. Kuriki, K., Koike, Y. and Okamoto, Y. (2002) Plastic optical fiber lasers and amplifiers containing lanthanide complexes. Chemical Reviews, 102, 2347-2356.http://dx.doi.org/10.1021/cr010309g

12. Bünzli, J.C.G. (2006) Benefiting from the unique properties of lanthanide ions. Accounts of Chemical Research, 39, 53-61. http://dx.doi.org/10.1021/ar0400894

13. Ple nik, C.E., Liu, S.M. and Shore, S.G. (2003) Lanthanide-transition-metal complexes: From ion pairs to extended arrays. Accounts of Chemical Research, 36, 499- 508.http://dx.doi.org/10.1021/ar010050o

14. Winpenny, R.E.P. (1998) The structures and magnetic properties of complexes containing 3dand 4f-metals. Chemical Society Reviews, 27, 447-452.http://dx.doi.org/10.1039/a827447z

15. Li, G.M., Akitsu, T., Sato, O. and Einaga, Y. (2003) Photoinduced Magnetization for Cyano-bridged 3d-4f Hetero-bimetallic Assembly $Nd(DMF)_4(m-CN)_5Fe(CN)_5 \cdot H_2O$ (DMF=N,N-dimethylformamide). Journal of the American Chemical Society, 125, 12396-12397. http://dx.doi.org/10.1021/ja037183k

16. Zhao, B., Chen, X.Y., Cheng, P., Liao, D.Z., Yan, S.P. and Jiang, Z.H. (2004) Coordination Polymers Containing 1D Channels as Selective Luminescent Probes. Journal of the American Chemical Society, 126, 15394-15395. http://dx.doi.org/10.1021/ja047141b

17. Qi, Y., Che, Y.X. and Zheng, J.M. (2008) A zinc(II) coordination polymer constructed from mixed-ligand 1,2- bis(2-(1H-imidazol-1-yl)ethoxy)ethane and 1,4-benzenedicarboxylic acid. CrystEngComm, 10, 1137-1139. http://dx.doi.org/10.1039/b801387j

18. Henninger, S.K., Habib, H.A. and Janiak, C. (2009) MOFs as adsorbents for low temperature heating and cooling

applications. Journal of the American Chemical Society, 131, 2776-2777. http://dx.doi.org/10.1021/ja808444z

19. Ma, L.F., Wang, L.Y., Wang, Y.Y., Du, M. and Wang, J.G. (2009) Synthesis, structures and properties of Mn(II) coordination frameworks based on R-isophthalate (R = $-CH_3$ or $-C(CH_3)_3$) and various dipyridyl-type co-ligands. CrystEngComm, 11, 109-117.http://dx.doi.org/10.1021/ja808444z

20. Habib, H.A., Sanchiz, J. and Janiak, C. (2009) Magnetic and luminescence Properties of Cu(II), $Cu(II)_4O_4$ core and Cd(II) mixed-ligand metal-organic frameworks constructed from 1,2-bis(1,2,4-triazol-4-yl)ethane and benzene-1,3,5-tricarboxylate. Inorganica Chimica Acta, 362, 2452-2460. http://dx.doi.org/10.1016/j.ica.2008.11.003

21. Habib, H.A., Hoffmann, A., Hoppe, H.A. and Janiak, C. (2009) Crystal structures and solid-state CPMAS ^{13}C NMR correlations in luminescent zinc(II) and cadmium(II) mixed-ligand coordination polymers constructed from 1,2-bis(1,2,4-triazol-4-yl)ethane and benzenedicarboxylate. Dalton Transactions, 1742-1751.http://dx.doi.org/10.1039/b812670d

22. Habib, H.A., Sanchiz, J. and Janiak, C. (2008) Mixedligand coordination polymers from 1,2-bis(1,2,4-triazol-4-yl) ethane and benzene-1,3,5-tricarboxylate: Trinuclear nickel or zinc secondary building units for three-dimensional networks with crystal-to-crystal transformation upon dehydration. Dalton Transactions, 1734-1744.http://dx.doi.org/10.1039/b715812b

23. Zhang, L.P., Yang, J., Ma, J.F., Jia, Z.F., Xie, Y.P. and Wei, G.H. (2008) A series of 2D and 3D metal-organic frameworks based on different polycarboxylate anions and a flexible 2,2'-bis(1H-imidazolyl)ether ligand. CrystEngComm, 10, 1410-1420.http://dx.doi.org/10.1039/b804578j

24. Wisser, B., Lu, Y. and Janiak, C. (2007). Chiral Coordination Polymers with Amino Acids: $^\infty_2$[Cu$_2$(μ-L-tryptophanato)$_2$(μ-

4,4'-bipyridine)(H$_2$O)$_2$](NO$_3$)$_2$. Zeitschrift für Anorganishe und Allgemeine Chemie, 633, 1189-1192.

25. Manna, S.C., Okamoto, K.I., Zangrando, E. and Chaudhuri, N.R. (2007) Conformational morphosis in azocalix arenes. CrystEngComm, 9, 199-122.

26. Chen, Z.F., Zhang, S.F., Luo, H.S., Abrahams, B.F. and Liang, H. (2007) Ni$_2$(R*COO)$_4$(H$_2$O)(4,4'-bipy)$_2$—A robust homochiral quartz-like network with large chiral channels. CrystEngComm, 9, 27-29. http://dx.doi.org/10.1039/b613047j

27. Pichon, A., Fierro, C.M., Nieuwenhuyzen, M. and James, S. (2007) A pillared-grid MOF with large pores based on the Cu$_2$(O$_2$CR)$_4$ paddle-wheel. CrystEngComm, 9, 449- 451. http://dx.doi.org/10.1039/b702176c

28. Pasán, J., Sanchiz, J., Lloret, F., Julve, M. and Ruiz-Pérez, C. (2007) Crystal engineering of 3-D coordination polymers by pillaring ferromagnetic copper(II)-methylmalonate layers. CrystEngComm, 9, 478-487. http://dx.doi.org/10.1039/b701788j

29. Stephenson, M.D. and Hardie, M.J. (2006) Coordination and hydrogen bonded network structures of Cu(II) with mixed ligands: A hybrid hydrogen bonded material, an infinite sandwich arrangement, and a 3-D net. Dalton Transactions, 3407-3417.http://dx.doi.org/10.1039/b600357e

30. Carballo, R, Covelo, B, Vázquez-López, E.M., GarcíaMartínez, E., Castiñeiras, A., Janiak, C. (2005) Zeitschrift für Anorganishe und Allgemeine Chemie, 631, 1929- 1931.

31. Yao, J., Lu, Z.D., Li, Y.Z., Lin, J.G., Duan, X.Y., Gao, S., Meng, Q.J. and Lu, C.S. (2008) Three-dimensional metal-organic frameworks constructed from bix and 1,2,4-benzenetricarboxylate. CrystEngComm, 10, 1379-1383. http://dx.doi.org/10.1039/b805263h

32. Wei, G.H., Yang, J., Ma, J.F., Liu, Y.Y., Li, S.L. and Zhang, L.P. (2008) Syntheses, structures and luminescent properties of zinc(II) and cadmium(II) coordination complexes based

on bis(imidazole) and different carboxylate ligands. Dalton Transactions, 3080-3092. http://dx.doi.org/10.1039/b716657e

33. Zhang, J.Y., Ma, Y., Cheng, A.L., Yue, Q., Sun, Q. and Gao, E.Q. (2008) A manganese(II) coordination polymer with mixed pyrimidine-2-carboxylate and oxalate bridges: synthesis, structure, and magnetism. Dalton Transactions, 2061-2066. http://dx.doi.org/10.1039/b717837a

34. Roh, S.G., Nah, M.K., Oh, J.B., Baek, N.S., Park, K.M. and Kim, H.K. (2005) Synthesis, crystal structure and luminescence properties of a saturated dimeric Er(III)- chelated complex based on benzoate and bipyridine ligands. Polyhedron, 24, 137-142. http://dx.doi.org/10.1016/j.poly.2004.10.014

35. Qin, C., Wang, X.L., Wang, E.B. and Su, Z.M. (2005) A series of three-dimensional lanthanide coordination polymers with rutile and unprecedented rutile-related topologies. Inorganic Chemistry, 44, 7122-7129. http://dx.doi.org/10.1021/ic050906b

36. Wan, Y.H., Jin, L.P. and Wang, K.Z. (2003) Hydrothermal synthesis and structural characterization of two novel lanthanide supramolecular coordination polymers with nano-chains. Journal of Molecular Structure, 649, 85-93. http://dx.doi.org/10.1016/S0022-2860(03)00021-8

37. Wan, Y., Zhang, L., Jin, L., Gao, S. and Lu, S. (2003) High-dimensional architectures from the self-assembly of lathanide ions with benzenedicarboxylates and 1,10- phenanthroline. Inorganic Chemistry, 42, 4985-4994. http://dx.doi.org/10.1021/ic034258c

38. Eddaoudi, M., Kim, J., Wachter, J.B., Chae, H.K., O'Keeffe, M. and Yaghi, O.M. (2001) Porous Metal-organic polyhedra: 25 Å cuboctahedron constructed from 12 $Cu_2(CO_2)_4$ paddle-wheel building blocks. Journal of the American Chemical Society, 123, 4368-4369. http://dx.doi.org/10.1021/ja0104352

39. Guo, X.D., Zhu, G.S., Sun, F.X., Li, Z.Y., Zhao, X.J., Li, X.T., Wang, H.C. and Qiu, S.L. (2006) Synthesis, structure, and

luminescent properties of microporous lanthanide metal-organic frameworks with inorganic rod-shaped building units. Inorganic Chemistry, 45, 2581-2587. http://dx.doi.org/10.1021/ic0518881

40. Thirumurugan, A. and Natarajan, S. (2004) Terephthalate bridge frameworks of Nd and Sm phthalates. Inorganic Chemistry Communications, 7, 395-399.http://dx.doi.org/10.1016/j.inoche.2003.12.023

41. Zhang, Z.H., Okamura, T.A., Hasegawa, Y., Kawaguchi, H., Kong, L.Y., Sun, Y.W. and Ueyama, N. (2005) Syntheses, structures, and luminescent and magnetic properties of novel three-dimensional lanthanide complexes with 1,3,5-benzenetriacetate. Inorganic Chemistry, 44, 6219-6227. http://dx.doi.org/10.1021/ic050453a

42. Cao, R., Sun, D., Liang, Y., Hong, M., Tatsumi, K. and Shi, Q. (2002) Syntheses and characterizations of threedimensional channel-like polymeric lanthanide complexes constructed by 1,2,4,5-benzenetetracarboxylic acid. Inorganic Chemistry, 41, 2087-2094.http://dx.doi.org/10.1021/ic0110124

43. Cañadillas-Delgado, L., Fabelo, Ó., Ruíz-Pérez, C., Delgado, F.S., Julve, M., Hernndez-Molina, M., Laz, M.M. and Lorenzo-Luis, P. (2006) Zeolite-like nanoporous gadolinium complexes incorporating alkaline cations. Crystal Growth and Design, 6, 87-93.http://dx.doi.org/10.1021/cg050170t

44. Tang S.F. and Mudring A.V. (2009) The missing link crystallized from the ionic liquid 1-ethyl-3-methylamonium tosylate: Bis-aqua-(p-toluenesulfonato-o)-europium(Ⅲ)- bis-toluenesulfonate dehydrate. Crystal Growth and Design, 9, 2549-2557.

45. Wang, Z., Bai, F.Y., Xing, Y.H., Xie, Y., Ge, M.F. and Niu, S.Y. (2010) Two New 3D Lanthanide Coordination Polymers with Benzenesulfonic and Adipic Acids: Synthesis, Structure and Luminescent Properties. Zeitschrift für Anorganishe und Allgemeine Chemie, 636, 1570- 1575.

46. Zhou, R.S., Ye, L., Ding, H., Song, J.F., Xu, X.Y. and Xu, J.Q. (2008) Sytheses, structures, luminescence, and magnetism

of four 3D lanthanide 5-sulfosalicylates, Journal of Solid State Chemistry, 181, 567-575. http://dx.doi.org/10.1016/j.jssc.2007.12.027

47. Lu, Z., Wen, L., Yao, J., Zhu, H. and Meng, Q. (2006) Two types of novel layer framework structures assembled from 5-sulfosalicylic acid and lanthanide ions. CrystEngComm, 8, 847-853. http://dx.doi.org/10.1039/b612147k

48. Ma, J.F., Li, J.Y., Zheng G.L. and Liu J.F. (2003) The first ladder structure containing three different squares: The structure of barium   3-carboxy-4-hydroxybenzenesulfonate.   Inorganic Chemistry Communications, 6, 581- 583. http://dx.doi.org/10.1016/S1387-7003(03)00044-3

49. Ma, J.F., Yang, J., Li, S.L., Song, S.Y., Zhang, H.J., Wan, H.S. and Yang, K.Y. (2005) Two Coordination polymers of Ag(I) with 5-sulfosalicylic acid. Crystal Growth and Design, 5, 807-812. http://dx.doi.org/10.1021/cg049723a

50. Gao, S., Zhu, Z.-B, Huo, L.-H. and Ng, S.W. (2005) Poly[[triaquabis(μ_4-3-carboxy-4-hydroxybenzenesulfonato) disilver(I)] monohydrate]. Acta Crystallographica, 61, m279-m281.http://dx.doi.org/10.1107/S1600536805000504

51. Gao, S., Zhu, Z.-B., Huo, L.-H. and Ng, S.W. (2005) Poly[di-μ-aqua-bis(μ_8-3-carboxylato-4-hydroxybenzenesulfonato) tetrasilver(I)]. Acta Crystallographica, 61, m282- m284.http://dx.doi.org/10.1107/S1600536805000516

52. Borkowski,   L.A.   and   Cahill,   C.L.   (2004)   Poly [[aquaneodymium(III)]-μ_3-decane-1,10-dicarboxylato-μ_3-9-carboxynonanecarboxylato]. Acta Crystallographica C, 60, m159-m161.http://dx.doi.org/10.1107/S0108270104003427

53. Cheetham, A.K., Férey, G. and Loiseau, T. (1999) OpenFramework   Inorganic   Materials.   Angewandte Chemie International Edition, 38, 3268-3292.http://dx.doi.org/10.1002/(SICI)1521-3773(19991115)38:22<3268::AID-ANIE3268>3.0.CO;2-U

54. Chesnut, D.J., Hagrman, D., Zapf, P.J., Hammond, R.P., LaDuca, R. Jr., Haushalter, R.C., Zubieta, J. (1999) Organic/

inorganic composite materials: The roles of organoamine ligands in the design of inorganic solids. Coordination Chemistry Reviews, 190-192, 737-769. http://dx.doi.org/10.1016/S0010-8545(99)00119-8

55. Bataille, T. and Louër, D. (2002) Two new diamine templated lanthanum sulfates, $La_2(H_2O)_2(C_4H_{12}N_2)(SO_4)_4$ and $La_2(H_2O)_2(C_2H_{10}N_2)_3(SO_4)_6.4H_2O$, with 3D and 2D crystal structures. Journal of Materials Chemistry, 12, 3487-3493. http://dx.doi.org/10.1039/b207212m

56. Rao, C.N.R., Behera, J.N. and Dan, M. (2006) Organically-templated metal sulfates, selenites and selenates. Chemical Society Reviews, 35, 375-387.http://dx.doi.org/10.1039/b510396g

57. Nakamoto, K., Fujita, J., Tanaka, S. and Kobayashi, M. (1957) Infrared spectra of metallic complexes. IV. Comparison of the infrared spectra of unidentate and bidentate metallic complexes. Journal of the American Chemical Society, 79, 4904-4908.http://dx.doi.org/10.1021/ja01575a020

58. Anbalagan, G., Mukundakumari, S., Murugesan, K.S. and Gunasekaran, S. (2007) Preparation, crystal structure and infrared spectroscopy of the new compound rubidium beryllium sulfate dihydrate, $Rb_2Be(SO_4)_2 \cdot 2H_2O$. Vibrational Spectroscopy, 44, 226-272.

59. Wang, T.W., Liu, D.S., Huang, C.C., Sui, Y., Huang, X.H., Chen, J.Z. and You, X.Z. (2010) Syntheses, crystal structures, and magnetic properties of two Mn(II) coordination polymers based on the 5-aminotetrazole ligand: Effect of sources of ligand on construction of topological networks. Crystal Growth and Design, 10, 3429-3435.http://dx.doi.org/10.1021/cg100127w

60. Bataille, T. and Louër, D. (2004) New linear and layered amine-templated lanthanum sufates. Journal of Solid State Chemistry, 177, 1235-1243.http://dx.doi.org/10.1016/j.jssc.2003.10.031

61. Zhu, Y., Sun, X., Zhu, D. and Xu, Y. (2009) Solvothermal synthesis, crystal structure and luminescence of the first

organic amine templated europium sulfate. Inorganica Chimica Acta, 362, 2565-2568. http://dx.doi.org/10.1016/j.ica.2008.11.015

62.   Xing, Y., Shi, Z., Li, G. and Pang, W. (2003) Hydrothermal synthesis and structure of $[C_2N_2H_{10}][La_2(H_2O)_4(SO_4)_4]$ $2H_2O$, a new organically templated rare earth sulfate with a layer structure. Dalton Transactions, 940-943. http://dx.doi.org/10.1039/b211076h

63.   Richardson, F.S. (1982) Terbium (III) and europium (III) ions as luminescent probes and stains for biomolecular systems. Chemical Reviews, 82, 541-552.http://dx.doi.org/10.1021/cr00051a004

64.   Allendorf, M.D., Bauer, C.A., Bhakta, R.K. and Houk, R.J.T. (2009) Luminescent metal-organic frameworks. Chemical Society Reviews, 38, 1330-1352.http://dx.doi.org/10.1039/b802352m

Integrated FCC Riser–Regenerator Dynamics Studied in a Fluid Catalytic Cracking Pilot Plant

G.M. Bollas[a], I.A. Vasalos[a], A.A. Lappas[b], D.K. Iatridis[b], S.S. Voutetakis[b], and S.A. Papadopoulou[c]

[a]Laboratory of Petrochemical Technology, Department of Chemical Engineering, Aristotle University of Thessaloniki, 57001 Thessaloniki, Greece

[b]Chemical Process Engineering Research Institute (CPERI), Centre for Research and Technology Hellas (CERTH), 6th km. Charilaou–Thermi Rd., GR-570 01 Thermi-Thessaloniki, Greece

[c]Automation Department, Alexander Technological Educational Institute of Thessaloniki, GR54101 Thessaloniki, Greece

ABSTRACT

In this paper a dynamic simulator of the fluid catalytic cracking (FCC) pilot plant, operating in the Chemical Process Engineering

Research Institute (CPERI, Thessaloniki, Greece), is presented. The operation of the pilot plant permits the execution of case studies for monitoring of the dynamic responses of the unit, by imposing substantial step changes in a number of the manipulated variables. The comparison between the dynamic behavior of the unit and that predicted by the simulator arise useful conclusions on both the similarities of the pilot plant to commercial units, along with the ability of the simulator to depict the main dynamic characteristics of the integrated system. The simulator predicts the feed conversion, coke yield and heat of catalytic reactions in the FCC riser on the basis of semi-empirical models developed in CPERI and simulates the regenerator according to the two-phase theory of fluidization, with a dilute phase model taking account of postcombustion reactions. The riser and regenerator temperature, the stripper and regenerator pressure drop and the composition of the regenerator flue gas are measured on line and are used for verification of the ability of the simulator to predict the dynamic transients between steady states in both open- and closed-loop unit operation. All the available process variables such as the reaction conversion, the coke yield, the carbon on regenerated catalyst and the catalyst circulation rate are used for the validation of the steady-state performance of the simulator. The comparison between the dynamic responses of the model and those of the pilot plant to step changes in the feed rate and preheat temperature reveals the ability of the simulator to accurately depict the complex pilot process dynamics in both open- and closed-loop operation. The dynamic simulator can serve as the basis for the development of a model-based control structure for the pilot plant, alongside its use as a tool for off-line process optimization studies.

INTRODUCTION

The dynamic simulation of the fluid catalytic cracking (FCC) process is a challenging research subject of high economic and environmental importance. Optimization of this complex process demands the development of accurate models capable of describing the process in detail. FCC technology is continuously evolving during the last half century, though the requirements for stable operation of commercial FCC units restrict the possibility of obtaining accurate models over an extensive operating range through experimentation. In industry the target is maximum capacity (i.e. profitability) and this bounds the application and observation of dynamic transients within particular and narrow operating windows. Thus, FCC pilot plants are often used for obtaining data useful for the simulation of commercial units under different operating conditions, feed properties and catalyst activities and selectivities. The operation of a pilot unit provides the ability to examine the process under steady feed and/or catalyst properties, in order to isolate their respective effects on the cracking reactions and develop correlations for each subset of process variables. One other asset of an FCC pilot plant is the potential to examine its dynamic behavior within a wide range of operating conditions, which enhances the possibility for the investigation of the process dynamics.

The research interest in dynamic simulation of the FCC process has been consecutively increasing during the last years. The first integrated FCC dynamic simulator published was developed by the research team of Amoco Oil Co (Ford et al., 1977). It included a dynamic model of the regenerator and a pseudo-steady-state riser model (Wollaston et al., 1975). Lee and Groves (1985) proposed a dynamic model which treated the riser as a pseudo-steady-state adiabatic plug flow reactor and the regenerator as a continuous stirred tank reactor with no dilute phase. Elnashaie et al. (1995) (Elnashaie and Elshishini, 1993) developed a dynamic model for an industrial-type IV FCC unit and investigated the sensitivity and stability of the system, which is of significant importance in this type of FCC unit. They used two-phase models for both the reactor and the regenerator with dynamic terms for the thermal behavior and the carbon mass balance throughout the entire unit. Lopez-Isunza (1992) presented a dynamic model with a 3-lump (Weekman and Nace, 1970) tubular reactor model of the riser and a moving bed bubble–emulsion dynamic model of the regenerator. McFarlane et al. (1993) presented a comprehensive model for the simulation of a type IV FCC unit, in which they included models of the blowers, U-bends, compressors, furnace and valves, in order to compute the pressure balance and catalyst circulation rate of this type of unit. In the regenerator model they included a dilute phase in account for postcombustion, but in the riser part they used oversimplified computations for the heat balance. In a very complete series of papers, Arbel et al., 1995a and Arbel et al., 1995b presented a model of the FCC in both steady-state and dynamic conditions, comprised by a 10-lump (Jacob et al., 1976) pseudo-steady-state riser model and a dynamic regenerator model, with detailed description of both full and partial combustion kinetics. They extensively studied the steady-state multiplicities of the FCC unit and the effect of combustion mode on the controllability of the unit. Ali and Rohani (1997) (Ali et al., 1997) presented a dynamic model, in which they developed analytical solutions of the differential model equations after adopting pseudo-steady-state assumptions. Their model neglected the freeboard region of the regenerator, which is important in the partial combustion mode. Secchi et al. (2001) presented a dynamic simulator for the UOP stacked FCC unit with a bubble–emulsion–freeboard model of the regenerator and a 10-lump (Jacob et al., 1976) dynamic model of the riser. They compared thedynamic performance of their simulator with experimental data of an industrial plant. Han et al. (2004) (Han and Chung, 2001a and Han and Chung, 2001b) presented a detailed dynamic simulator of the FCC process, in which they included models of the catalyst liftlines, stripper, feed preheater and cyclones. They applied a distributed parameter 4-lump model for the riser reactor and a two-regime, two-phase model for the regenerator. The UOP-type FCC unit was also simulated by Cristea et al. (2003). The simulator of Cristea et al. was based on the model of McFarlane et al. (1993) including models of the feed preheater, main fractionator, air blower and wet gas compressor, which were all implemented in a model predictive control (MPC) algorithm. Recently, Hernandez-Barajas et al. (2006) presented another dynamic simulator of the FCC unit, with a detailed pressure balance of the unit and studied the multiplicity of steady states. Generally, the models presented in the open literature deal with commercial FCC units. That means that they are of high economic importance, but they are bounded within the narrow process window of the refineries with very low ability of model validation.

In this paper the development and verification of a dynamic simulator, on the basis of steady-state and dynamic experimental data of the FCC pilot plant of Chemical Process Engineering Research Institute (CPERI), will be presented. The term "dynamic experiments" is used to describe experiments, in which a step change is imposed to a manipulated process variable, while recording the transient of a number of process variables from the original steady state to the final steady state the system will eventually reach. The development of the dynamic simulator of the pilot plant serves two main goals: (a) the study of the dynamic behavior of the pilot process that includes the validation of the model performance against steady-state and dynamic unit responses, the identification of the process dependencies and uncertainties and the performance of experimental case studies to examine the similarities of the pilot plant with commercial units; and (b) the use of the simulator for the development of a model-based optimizer and control scheme for the entire unit. This paper deals with the former of the aforementioned goals and examines the ability of the simulator to provide an accurate representation of the unit qualitatively and quantitatively, as well as the ability of both the pilot plant and the simulator to depict the main dynamic characteristics of a typical commercial FCC process.

EXPERIMENTAL SETUP

The FCC pilot plant of CPERI (Fig. 1) operates in a fully circulating mode and consists of a riser reactor, a fluid bed regenerator, a stripper and a liftline (Vasalos et al., 1996). The riser reactor operates at pseudo-isothermal plug flow conditions and consists of a large-diameter bottom region (mixing zone) (26 mm i.d., 0.3 m height) and a smaller-diameter (7 mm i.d., 1.465 m height) top region connected by a conical-shaped region of 0.05 m height. At the reactor bottom, the gas–oil contacts the hot catalyst (which flows from the regenerator) and evaporates, while the catalyst is kept in a fluidized state by means of nitrogen flow. The cracking product from the riser top enters the stripper vessel for the separation (stripping) of vapors from the catalyst. The cracked gas flows through a heat exchanger for condensation of the heavier compounds. Thereupon, the liquid and gaseous products are separated in a top cooled stabilizer column operating in 2.2 atm. The mixture of gasoline, light cycle oil and heavy cycle oil is obtained through the bottom of the stabilizer. The yield to liquid products is measured with the ASTM D-2887 simulated distillation method. The stripped catalyst flows through a liftline to the regenerator, where the carbon, deposited on the catalyst surface, is burned off. The regenerator consists of two main sections. A small-diameter bottom section (77.92 mm i.d., 0.715 m height) and a larger-diameter top section (254.6 mm i.d., 0.64 m height) connected by a conical-shaped section of 0.205 m height. A standpipe at the bottom of the regenerator leads the regenerated catalyst back to the riser bottom to continue the operation loop. Two slide valves one at the exit of the regenerator standpipe and one at the exit of the stripper standpipe regulate the catalyst circulation throughout the unit. The regenerator standpipe slide valve controls the catalyst circulation to satisfy a predetermined set point for the riser temperature, while the stripper slide valve operates for constant stripper level (i.e. stripping volume). Two wet test meters and two gas chromatographers measure the volumetric flow rates and the molar composition of the flue and cracked gas, respectively. An on-line oxygen analyzer monitors the excess of oxygen to ensure thorough catalyst regeneration. The process control system of the unit is based on a special industrial computer control system. The system is coordinated with the iFIX S/W by Intellution GE. The control system collects the values of the inputs and drives the output signals as well as maintains a digital record of the signals. The process pressure control valves and the power to electrical heaters are controlled by numerous algorithmic PID controllers. The operation of electrical heaters is responsible for establishing isothermal riser and adiabatic regenerator operating profiles.

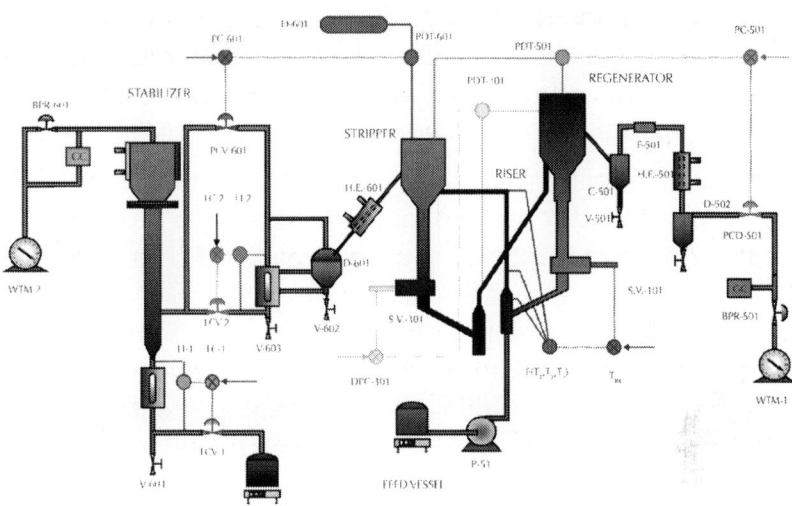

Figure 1: Schematic diagram of the FCC pilot plant of CPERI.

MODEL DESCRIPTION

The simulator of the pilot plant includes three main sections: a pseudo-steady-state model of the riser reactor, a dynamic model of the regenerator and a set of dynamic and pseudo-steady-state models of the stripper, the regenerator standpipe, the liftline and the slide valves. For the specific case of the pilot plant, the dynamic effects of the riser, the cyclones, the liftline and the regenerator standpipe were neglected, because their operation has significantly lower impact on the process dynamics, compared to the two large vessels of the pilot plant, the stripper and the regenerator. In the pilot plant, or in a typical commercial unit, the behavior of the regenerator dominates on both the dynamic and the steady-state behavior of the integrated unit (Arbel et al., 1995a). The riser residence times are much shorter compared to those of the regenerator, hence one can at any instance describe the riser reactor by a set of pseudo-steady-state equations, which simplifies the dynamic analysis. The main impact of the riser operation on both the dynamic and steady-state behavior of the integrated system is on the coke production and on the heat consumption. Thus, the accurate prediction of pseudo-steady-state conversion, coke yield and heat of cracking and vaporization is significant, when describing the effect of riser in the integrated dynamic system.

The pseudo-steady-state and dynamic submodels that constitute the dynamic simulator of the pilot plant have been presented in the literature (Bollas et al., 2002, Bollas et al., 2004, Faltsi-Saravelou and Vasalos, 1991 and Faltsi-Saravelou et al., 1991) and will be briefly adduced in this section. A kinetic–hydrodynamic model was developed for the simulation of the pilot riser reactor in pseudo-steady-state conditions (Bollas et al., 2002). The catalyst hold-up and residence time in the reactor were calculated on the basis of empirical hydrodynamic correlations and the gas–oil conversion and coke yield were predicted through a Blanding-type (Blanding, 1953) kinetic model. The prediction of gas–oil conversion and coke yield were the only 2 lumps of the riser submodel, essential for inclusion in the complete simulator, which reduced the necessity of using a more detailed lumped model. The primary reason for the latter is that the control problem of the pilot process concerns only the feed conversion. The pilot plant is mainly operated for catalyst/feedstock benchmarking; therefore, it is necessary to operate in constant feed conversion and riser temperature, as to observe the effect of the examined catalyst or feedstock on the product selectivity. The effect of feedstock properties on gas–oil conversion and coke yield was expressed through semi-empirical correlations developed on the basis of experiments performed with constant catalyst and a variety of feedstocks (Bollas et al., 2004). The effect of catalyst type was expressed through a "catalyst index" (Bollas et al., 2004). The model of the regenerator was based on the two-phase theory (Davidson et al., 1985), in which the gas–solids flow is assumed to follow the bubbling bed regime, consisting of two zones: a dense zone at the regenerator bottom comprising of a bubble and an emulsion phase, and a dilute zone at the regenerator top called the freeboard. The model equations were grouped into two main modules that serve for the two main sections of the unit, the riser and the regenerator. A third module was used for the simulation of the stripper and the slide valves, the liftline and the standpipe.

Simulation of Riser Reactor

The pseudo-steady-state model of the FCC riser reactor (Bollas et al., 2002 and Bollas et al., 2004) was developed on the basis of the following assumptions:

- The aggregate effects of operating conditions, feed properties and catalyst type on the cracking reactions are simulated by the product of their discrete functions,
- The riser reactor is assumed to run in concurrent plug flow of gas and solids at pseudo-isothermal conditions, which is accurate for the case of the pilot plant,
- Second-order rate apparent kinetics are applied for gas–oil conversion (x),
- Catalytic coke (cx) deposition parallels catalyst deactivation (Voorhies, 1945).

On the basis of these assumptions and after integration and rearrangement of the corresponding spatial equations, Eqs. (1) And (2) were formulated:

$$\frac{x}{100-x} = C(\text{catalyst type}) F(\text{feed quality}) \frac{k_x}{\text{WHSV}}$$
$$\times \exp\left(\frac{-E_x}{R_g T_{RX}}\right) t_{C:RS}^{n_x}, \tag{1}$$

$$c_x = C_c(\text{catalyst type}) F_c(\text{feed quality}) \frac{k_c}{\text{WHSV}}$$
$$\times \exp\left(\frac{-E_c}{R_g T_{RX}}\right) t_{C:RS}^{n_c}. \tag{2}$$

In Eqs. (1) And (2) C, $_{Cc}$ and F, $_{Fc}$ describe the effects of catalyst and feedstock properties on the conversion (x) and coke yield ($_{cx}$), respectively. The adjustable parameters (kx, kc, Ex, Ec, nx, nc) of Eqs. (1) and (2) were estimated on the basis of a data set of steady-state pilot experiments performed with constant feed and catalyst quality, in a great range of space velocities (weight hourly space velocity (WHSV)) and catalyst residence times (tC:RS) and at two different reactor temperatures (TRX) (Bollas et al., 2002). A large database of experiments with different feedstocks and catalysts was used for the development of models of the effect of feedstock quality and the assignment of "catalyst indices" that are used in Eqs. (1) and (2). Moreover, a model was developed for the prediction of the non-catalytic coke yield, as a function of feedstock properties. The methodology of the simulation of the effect of feed and catalyst quality on conversion and coke yield was described elsewhere (Bollas et al., 2004). The values of the parameters of Eqs. (1) And (2) are given in Table 1.

Table 1: Parameters of the model of the pilot riser

Kinetic model		Heat of cracking	
kx	200.04	a1	3.49
E_x (kcal mol^{-1})	8.9	a2	18.6
nx	-0.78	a3	0.021
kc	1.283	b1	−9.50
E_c (kcal mol^{-1})	0.9	b2	53.4
nc	−0.90	b3	0.044

Finally, a pseudo-steady-state heat balance of the riser reactor was performed. The main contributors to the overall enthalpy balance in an FCC plant are: (a) the enthalpy of cracking Hcrack, (b) the enthalpy of vaporization of the gas–oil feedstock, and (c) the enthalpy content of various process streams (gas–oil, catalyst, cracked products, inerts). An empirical correlation was developed to estimate the heat of cracking in the riser reactor. This correlation was based on experiments performed at different temperatures, using various feedstocks and at different conversion levels. The final correlation estimates the heat of cracking as a function of conversion, riser temperature and gas–oil molecular weight, as shown in Eq. (3). The proposed empirical correlation was verified with pilot plant steady-state experiments with different feedstocks and operating conditions with satisfying results.

$$\Delta H_{crack} = \ln\left(\frac{x}{100-x}\right)(a_1 T_{RX} + a_2 T_{RX}^2 + a_3 MW_F)$$
$$+ (b_1 T_{RX} + b_2 T_{RX}^2 + b_3 MW_F). \tag{3}$$

The enthalpy content of gas–oil vapors was estimated by integration of the empirical correlation of Kesler and Lee (1976). The values of the Watson characterization factor (KW) and molecular weight (MWF) of the feedstock that are required in Eq. (3) and in the Kesler–Lee correlation were 11.58 and 392.5, respectively. The values of the parameters of Eq. (3) are presented in Table 1.

Simulation of Regenerator

The simulator of the regenerator was based on the model of Faltsi-Saravelou and Vasalos (1991). Many modifications were made to this model for its application on the pilot plant. The physical model of the regenerator includes two zones: (a) the dense bed and (b) the freeboard. The dense bed consists of a bubble and an emulsion phase, while the freeboard contains the entrained catalyst particles that are recycled to the emulsion phase of the dense zone. The assumptions made for the simulation of each phase (Faltsi-Saravelou and Vasalos, 1991 and Faltsi-Saravelou et al., 1991) are:

- The bubble phase is free of catalyst particles,
- Plug flow regime is assumed for the bubble phase,
- The emulsion phase gas and catalyst particles are assumed fully mixed,
- The freeboard is modeled as an ideal plug flow reactor,

- The catalyst particles are hydrodynamically represented by their average size, density and porosity, while the particle size distribution is used for the emulsion to freeboard entrainment rate calculation,
- Diffusion in the catalyst particles is neglected, due to their small size,
- Due to the high temperatures in the FCC regenerator, the ideal gas law is valid,
- The fluidized bed reactor is adiabatic.

The velocity of the gas flowing through the emulsion phase was assumed to be equal to the minimum bubbling velocity, which is a consistent assumption for Geldart group A particles (Geldart, 1973) (to which category FCC catalysts typically belong). The clouds and wakes around the bubbles were assumed to have zero volume. This assumption is valid for high ratios of superficial gas velocity over minimum fluidization velocity, which is typical for operations of group A particles. The bubbles were assumed to grow in size with bed height, while the variation of the fluidizing gas density and superficial velocity, due to axial temperature gradients and gas molar rate changes, was also taken into account.

The dense bed of the regenerator was simulated as a pseudo-steady-state PFR (bubble phase) in parallel to a dynamic CSTR (emulsion phase). The dense bed volume was calculated on the basis of the overall regenerator dynamics:

$$\frac{dV_{D:RG}}{dt} = \frac{\dot{W}_{C:RG}^{(l_D=1)} - \dot{W}_{C:RG}^{(l_D=0)} + \dot{W}_{C:CY}^{(l_F=1)} - \dot{W}_{C:CY}^{(l_F=0)}}{\rho_p(1 - \varepsilon_e)f_e}. \tag{4}$$

In the emulsion phase the material balance equations were formulated separately for gas and solid components and included the terms of accumulation, input and output rates, bubble–emulsion interchange and reaction generation or consumption as shown in

$$f_e \varepsilon_e \frac{dc_{ie}}{dt} = \frac{\dot{W}_{ge}^{(l_D=0)}}{\rho_{ge}} \frac{c_{ie}^{(l_D=0)} - c_{ie}}{V_{D:RG}}$$

$$+ \int_0^1 K_{Mi} \, dl_D + f_e \varepsilon_e \sum_j^{homo} a_{ij} K_{Rje}$$

$$+ f_e(1 - \varepsilon_e) \sum_j^{hete} a_{ij} K_{Rje}, \tag{5}$$

$$f_e(1 - \varepsilon_e) \frac{dc_{ie}}{dt} = \frac{\dot{W}_{C:RG}^{(l_D=1)}}{\rho_p} \frac{c_{ie}^{(l_D=1)} - c_{ie}}{V_{D:RG}}$$

$$+ \frac{\dot{W}_{C:CY}^{(l_F=1)}}{\rho_p} \frac{c_{if}^{(l_F=1)} - c_{ie}}{V_{D:RG}}$$

$$+ f_e(1 - \varepsilon_e) \sum_j^{hete} a_{ij} K_{Rje}. \tag{6}$$

The energy balance equation in the emulsion phase is given by

$$\left(f_e(1 - \varepsilon_e) \sum_i^{solids} c_{ie} c_{pie} + f_e \varepsilon_e \sum_i^{gas} c_{ie} c_{pie} \right) \frac{d(V_{D:RG} T_e)}{dt}$$

$$= Q_{C:RG}^{(l_D=1)} - Q_{C:RG}^{(l_D=0)} + Q_{C:CY}^{(l_F=1)} - Q_{C:CY}^{(l_F=0)} + Q_{ge}^{(l_D=0)}$$

$$- Q_{ge}^{(l_D=1)} - Q_{loss} + V_{D:RG} \int_0^1 K_H \, dl_D$$

$$+ f_e \varepsilon_e V_{D:RG} \sum_j^{homo} (-\Delta H_{Rj}) K_{Rje}$$

$$+ f_e(1 - \varepsilon_e) V_{D:RG} \sum_j^{hete} (-\Delta H_{Rj}) K_{Rje}. \tag{7}$$

The material balance for gas components in the bubble phase is

$$\frac{1}{V_{D:RG}} \frac{dF_{ib}}{dl_D} = -K_{Mi} + f_b \sum_{j}^{homo} a_{ij} K_{Rjb}.$$

(8)

The energy balance in the bubble phase is given by

$$\frac{1}{V_{D:RG}} \frac{dQ_b}{dl_D} = -K_H + f_b \sum_{j}^{homo} (-\Delta H_{Rj}) K_{Rjb},$$

(9)

The superficial bubble gas velocity for the dimensionless fraction of dense bed height$_{lD}$ is evaluated by differentiating the ideal gas law in terms of the bubble enthalpy rate term:

$$\frac{du_{gb}}{dl_D} = \frac{R_g}{A_{D:RG} P_{D:RG} \bar{c} \bar{p}_{gb}} \frac{dQ_b}{dl_D}.$$

(10)

The combined bubble to emulsion gas interchange coefficients are evaluated by

$$\frac{f_b}{K_{ti}} = \frac{1}{k_{bci}} + \frac{1}{k_{cei}}.$$

(11)

For the evaluation of the bubble-cloud (kbci) and cloud-emulsion (kcei) gas interchange coefficients the expressions proposed by Kunii and Levenspiel (1977) were adopted. The same method was used for the estimation of the heat interchange coefficient (Ht) (Kunii and Levenspiel, 1977).

The bubble–emulsion mass interchange KMi and the heat interchange KH and the emulsion fraction fe are evaluated by Eqs. (12)– (14), respectively:

$$K_{Mi} = K_{ti} \left(\frac{F_{ib}}{u_{gb} A_{D:RG}} - c_{ie} \right),$$

(12)

KH=Ht(Tb-Te),

(13)

$$f_e = \int_0^1 (1 - f_b) \, dl_D.$$

(14)

The freeboard is simulated as an ideal two-phase PFR. The

material balances of the gas and solid components in the freeboard are shown in Eqs. (15) and (16), respectively:

$$\frac{1}{V_{F:RG}}\frac{dF_{if}}{dl_F} = \varepsilon_f \sum_j^{\text{homo}} \alpha_{ij} K_{Rjf} + (1 - \varepsilon_f) \sum_j^{\text{hete}} \alpha_{ij} K_{Rjf},$$
(15)

$$\frac{1}{V_{F:RG}}\frac{dF_{if}}{dl_F} = (1 - \varepsilon_f) \sum_j^{\text{hete}} \alpha_{ij} K_{Rjf}.$$
(16)

The energy balance for the freeboard is

$$\frac{1}{V_{F:RG}}\frac{dQ_f}{dl_F} = \varepsilon_f \sum_j^{\text{homo}} (-\Delta H_{Rj}) K_{Rjf}$$

$$+ (1 - \varepsilon_f) \sum_j^{\text{hete}} (-\Delta H_{Rj}) K_{Rjf}.$$
(17)

The ideal gas law is differentiated in terms of the gas enthalpy rate to evaluate the gas superficial velocity:

$$\frac{du_{gf}}{dl_F} = \frac{R_g}{A_{F:RG} P_{F:RG} \overline{cp}_{gf}} \frac{dQ_{gf}}{dl_F}.$$
(18)

The derivative of the enthalpy of the gas phase is obtained by Eq. (19), assuming that the heat capacity of the components is constant at each integration step:

$$\frac{dQ_{gf}}{dl_F} = \frac{Q_{gf}}{Q_f} \frac{dQ_f}{dl_F}.$$
(19)

The chemical species considered to be involved in the reaction scheme of the regenerator are categorized to gas and solid components. A short description of the kinetics of the heterogeneous and homogeneous reactions is given in (R1)–(R6):

The intrinsic carbon combustion on the catalyst surface corresponds to a couple of reactions producing CO and CO_2 with second-order kinetics:

$$C + \tfrac{1}{2}O_2 \xrightarrow{K_1} CO, \quad r_1 = K_1[C][O_2],$$
(R1)

$$C + O_2 \xrightarrow{K_2} CO_2, \quad r_2 = K_2[C][O_2]. \tag{R2}$$

The homogeneous CO oxidation in the gas phase at which the water acts catalytically is

$$CO + \tfrac{1}{2}O_2 \xrightarrow{K_3} CO_2, \quad r_3 = K_3[O_2]^{0.5}[CO][H_2O]^{0.5}. \tag{R3}$$

The catalytic CO oxidation at which part of the CO produced on the catalyst site is catalytically oxidized on the catalyst itself or on an oxidation promoter is

$$CO + \tfrac{1}{2}O_2 \xrightarrow{K_4} CO_2, \quad r_4 = K_4[CO]. \tag{R4}$$

The hydrogen combustion on the catalyst surface which produces a significant amount of heat, is

$$2H + \tfrac{1}{2}O_2 \xrightarrow{K_5} H_2O, \quad r_5 = K_5[H][O_2]. \tag{R5}$$

The coke sulfur combustion on the catalyst, which produces mainly SO_2, is

$$S + O_2 \xrightarrow{K_6} SO_2, \quad r_6 = K_6[S][O_2]. \tag{R6}$$

The reaction of C and CO_2 on the catalyst site producing CO is neglected, as it occurs at a very low rate. The parameters of the kinetic expressions (R1)–(R6) are presented in Table 2.

Table 2: Parameters of the model of the regenerator

Frequency factor		Activation energy		Reference
		$(\mathrm{kcal\,mol}^{-1})$		
K1+K2	1.4E05	E1	29.9	Morley and De Lasa (1987)
K1/K2	2.5E03	E2	12.4	Arthur (1951)
K3	1.3E03	E3	30.0	Howard et al. (1973)
K4	3.5E03	E4	13.8	Tone et al. (1972)
K5	3.3E07	E5	37.7	Wang et al. (1986)
K6	1.4E05	E6	29.9	Faltsi-Saravelou et al. (1991)

Simulation of Stripper, Regenerator Standpipe and Liftline

The stripper was simulated as a perfectly mixed reactor in minimum fluidization conditions. The stripping volume and the material balance for the solid components were expressed through

$$\frac{dV_{D:ST}}{dt} = \frac{\dot{W}_{C:ST}^{(l_D=1)} - \dot{W}_{C:ST}^{(l_D=0)}}{\rho_p(1 - \varepsilon_{mf})}, \tag{20}$$

$$\frac{dc_{i:ST}}{dt} = \frac{\dot{W}_{C:ST}^{(l_D=1)}(c_{i:ST}^{(l_D=1)} - c_{i:ST})}{\rho_p(1 - \varepsilon_{mf})V_{D:ST}}. \tag{21}$$

The stripping efficiency of the pilot stripper was assumed to be 100%, as the stripper volume and the stripping steam flow are adequately large for the pilot riser capacity. In the pilot plant the temperature of the catalyst at the stripper dense bed is regulated by electrical heaters that operate for the achievement of a set point value.

The only effect the liftline and the regenerator standpipe have on the integrated pilot plant is the change of temperature of the catalyst stream due to heat loss and the transport lag of the catalyst. The temperature of the catalyst stream at the exit of the regenerator standpipe (riser entrance) and at the exit of the liftline (regenerator entrance) was calculated by modeling the heat loss throughout their height with

$$\frac{dT}{dl} = (T_w - T)\frac{U_w\pi D_w L_w}{\dot{W}_C cp_C}. \tag{22}$$

Model Structure—Initial and Boundary Conditions

The dynamic material and energy balance equations form a system of integro-differential equations that is solved following an iterative procedure commencing from the initial and boundary conditions of the system, as shown in Fig. 2. The common case is that the simulator is used to study the transient from a simulated steady state of the unit to a new one, when a step change is imposed to one or more of the manipulated variables. Otherwise, the system variables receive the starting values of a "guess" steady state, estimated on the basis of the assumption of plug flow conditions throughout the regenerator. Thereafter, the system is solved until convergence to a valid steady state. For time zero the emulsion variables ($C_{ie}^{i=0}$, $V_{D:RG}^{t=0}$, $T_e^{t=0}$ and all other time dependent variables shown in Fig. 2) hold the values of the initial steady state. The superficial gas velocity at the regenerator entrance ($u_{g:RG}^{t_b=0}$) is calculated using the ideal gas law for combustion air flow rate ($\dot{W}_{g:RG}^{t_b=0}$) at temperature ($T_b^{t_b=0}$) equal to the air preheat temperature and pressure ($P_{D:RG}^{t_b=0}$) equal to the regenerator bottom pressure. The superficial gas velocity in the bubbles at the entrance of the regenerator is then calculated by Eq. (23), where $u_{ge}^{(t)}$ is the superficial gas velocity in emulsion at real time t ($t=0$ for the initial steady state):

$$u_{gb}^{(t_D=0)} = u_{g:RG}^{(t_D=0)} - u_{ge}^{(t)}.$$

$$(23)$$

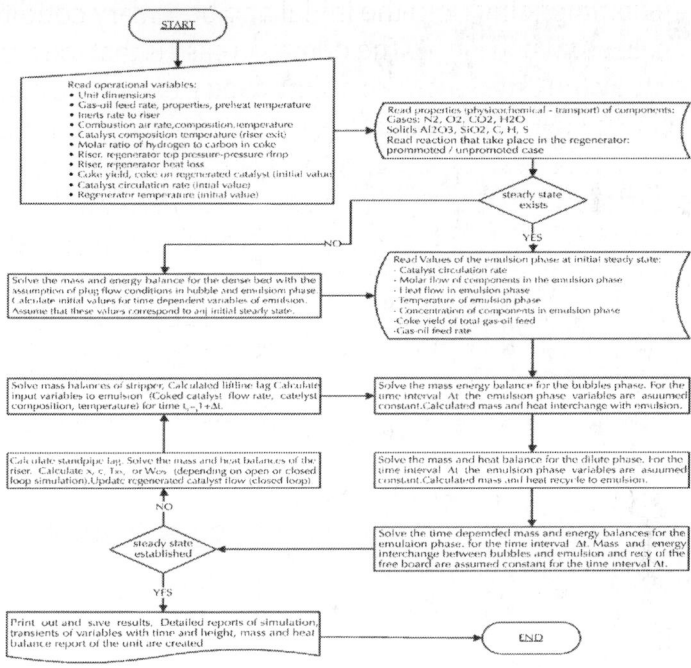

Figure 2: Logical scheme of the dynamic simulator of the FCC pilot plant of CPERI.

For the dilute phase, the boundary conditions at the dimensionless height $_{lF}=0$ (end of the dense zone entrance to the freeboard region) are

$$F_{if}^{(l_F=0)} = F_{ib}^{(l_D=1)} + c_{ie}^{(t)} u_{ge} A_{D:RG},$$

(24)

$$F_{if}^{(l_F=0)} = c_{ie}^{(t)} u_{sf}^{(l_F=0)} A_{F:RG},$$

(25)

$$Q_f^{(l_F=0)} = Q_b^{(l_D=1)} + \left(\sum_i^{\text{solids}} c_{ie}^{(t)} u_{sf}^{(l_F=0)} A_{F:RG} cp_{ie} \right.$$

$$\left. + \sum_i^{\text{gas}} c_{ie}^{(t)} u_{ge}^{(l_D=1)} A_{D:RG} cp_{ie} \right) (T_e - T_r),$$

(26)

$$u_{gf}^{(l_F=0)} = (u_{gb}^{(l_D=1)} + u_{ge}^{(t)}) \frac{A_{D:RG}}{A_{F:RG}}.$$

(27)

The catalyst with concentration $C_{ie}^{(t)}$ enters the riser with rate that is determined by the slide valve at the end of the regenerator standpipe (Eq. (28)) after time lag given by

$$\dot{W}_{C:RS} = k_{SV_1} \left(\frac{A_{SP}^2 A_{t:SV_1}^2}{A_{SP}^2 - A_{t:SV_1}^2} \right)^{0.5} (2\rho_p(1 - \varepsilon_b)(P_{RG}^{(l_F=1)}$$

$$+ \Delta P_{RG} + \Delta P_{SP} - P_{RS}^{(l_F=1)} - \Delta P_{RS}))^{0.5},$$

(28)

$$t_{dead}^{(RG \to RS)} = \frac{V_{SP}\rho_p(1 - \varepsilon_b)}{\dot{W}_{C:RG}^{(l_D=0)}}.$$

(29)

The same formulation is used to calculate the catalyst mass flow rate entering the regenerator after time lag determined by the residence time of the catalyst in the liftline:

$$\dot{W}_{C:RG}^{(l_D=1)} = k_{SV_2} \left(\frac{A_{ST}^2 A_{t:SV_2}^2}{A_{ST}^2 - A_{t:SV_2}^2} \right)^{0.5} (2\rho_p(1 - \varepsilon_{mf})(P_{ST}^{(l_F=1)}$$

$$+ \Delta P_{ST} - P_{RG}^{(l_F=1)} - \Delta P_{LL}))^{0.5},$$

(30)

$$t_{dead}^{(SP \to RG)} = \frac{V_{LL}\rho_p(1 - \varepsilon_{mf})}{\dot{W}_{C:ST}^{(l_D=0)}}.$$

(31)

The industrial practice for profitable and constant operation of the FCC unit is to control the riser exit temperature. The automatic control of the reactor temperature was included in the simulator with a routine that adjusts the catalyst circulation rate for constant riser temperature. Using the pseudo-steady-state model of the riser, the catalyst circulation rate is regulated by solving the system commencing from the heat balance (Eq. (3)) and the conversion and coke yield equations (Eqs. (1), (2)) simultaneously, at each solution cycle, as shown in Fig. 3. In Fig. 3 the regenerator temperature, the riser temperature, the coke on regenerated catalyst, the rate and quality of the feedstock and the inerts rate are used for the calculation of the catalyst circulation rate that redeems the mass and energy balances in the riser. Based on the new calculated value for the catalyst circulation rate and the estimate of coke production, the stripper entrance variables and the regenerator exit flow are updated and the loop continues, until the convergence that declares that steady state is achieved.

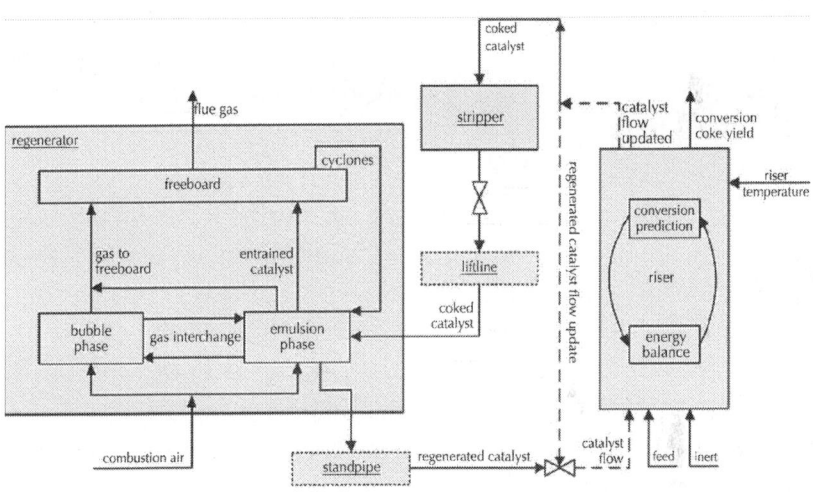

Figure 3: Structure of solution sequence of the integrated FCC simulator.

MODEL APPLICATION TO THE PILOT PLANT OF CPERI

The simulator was used to predict the behavior of the pilot plant when step changes in the flow rate and preheat temperature of the feed were applied. In the pilot plant both open- and closed-loop experiments were carried out. Open-loop experiments were called the pilot experiments, in which the reactor temperature control loop that operates on catalyst circulation rate was opened. In the closed-loop experiments the riser temperature was controlled by catalyst circulation rate to satisfy a predetermined set point. Accordingly, the part of the simulator that adjusts the catalyst circulation rate for constant riser temperature was set active or inactive. This resulted in four simulated pilot plant experiments:

- dynamic open-loop operation when decreasing the feed flow rate by 15%,
- dynamic closed-loop operation when decreasing the feed flow rate by15%,
- dynamic open-loop operation when increasing the feed preheat temperature by 130%,
- dynamic closed-loop operation when increasing the feed preheat temperature by 130%.

The open-loop experiments were used for validation of the mass and energy balances accuracy and for verification of the overall structure of the integrated model (assumption of pseudo-steady-state operation of the riser, iterative procedure of convergence, etc.). In the open-loop experiments the actions of the controller of riser temperature by catalyst circulation rate do not interfere with the process dynamics and the net dynamic responses of the unit can be observed. The closed-loop experiments were performed to examine the effect of the controller of riser temperature by catalyst circulation rate and were compared with the dynamic model responses, with the algorithmic adjustment of catalyst circulation rate being active. All experiments were performed with constant feedstock and catalyst, the properties of which are presented in Table 3.

Table 3: Bulk properties of the feedstock and the catalyst used in the experiments examined

Feedstock properties		Catalyst properties	
Gravity (API)	18.9	Bulk density Kgm^{-3}	900
Refractive index (at $20\,^{\circ}C$)	1.5226	Mean particle diameter (μm)	75
Sulfur (wt%)	2.58	Al_2O_3 (wt%)	39.1
Nitrogen (wt%)	0.13	SiO_2 (wt%)	59.6
Carbon (wt%)	85.3	Re_2O_3 (wt%)	0.65
Con. carbon residue (wt%)	0.36		
		Particle size distribution	
TBP distillation (°C)		Fraction (wt%)	Size (μm)
IBP	303.6		
10%	379.2	0	0
30%	422.9	10	53
50%	454.4	50	83
70%	483.1	80	104
90%	524.2	90	125
FBP	551.6	100	140

It should be noted that the pilot plant shows large deviation (low response) from the ideal instantaneous step change ordered. The 15% decrease in feed rate was established in the real time operation of the pilot plant within a 5 min period; while an average of 80 min was required for the 130% increase of the feed preheat temperature. At this point, the development of the dynamic simulator focuses on the study of the integrated riser–regenerator system; hence it does not include models for the dynamics of the feed mass flow meter and heater. The transients in the imposition of the changes in question were, however, recorded by the control system of the pilot plant. Thus, it was possible to reproduce the transients of change in the manipulated variables in five consecutive representative steps, as shown in Fig. 4, Fig. 5, Fig. 6 and Fig. 7. These representative step changes are the very fact of what was called "step change" in this study and these were entered to the simulator.

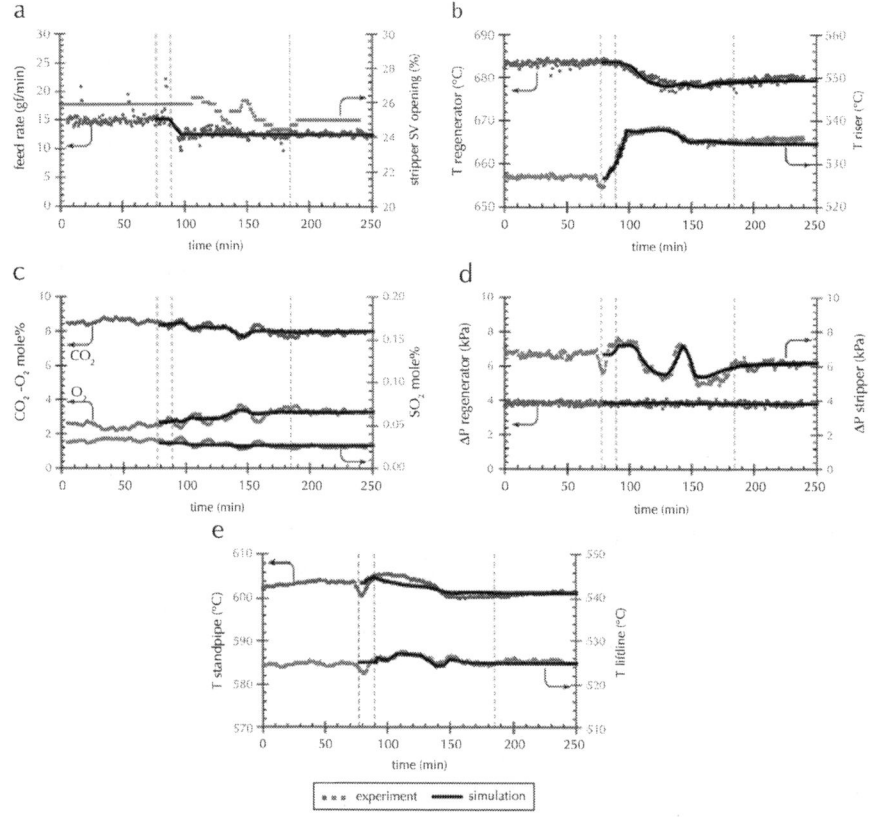

Figure 4: Open-loop responses of pilot plant and simulator for a 15% decrease in feed rate.

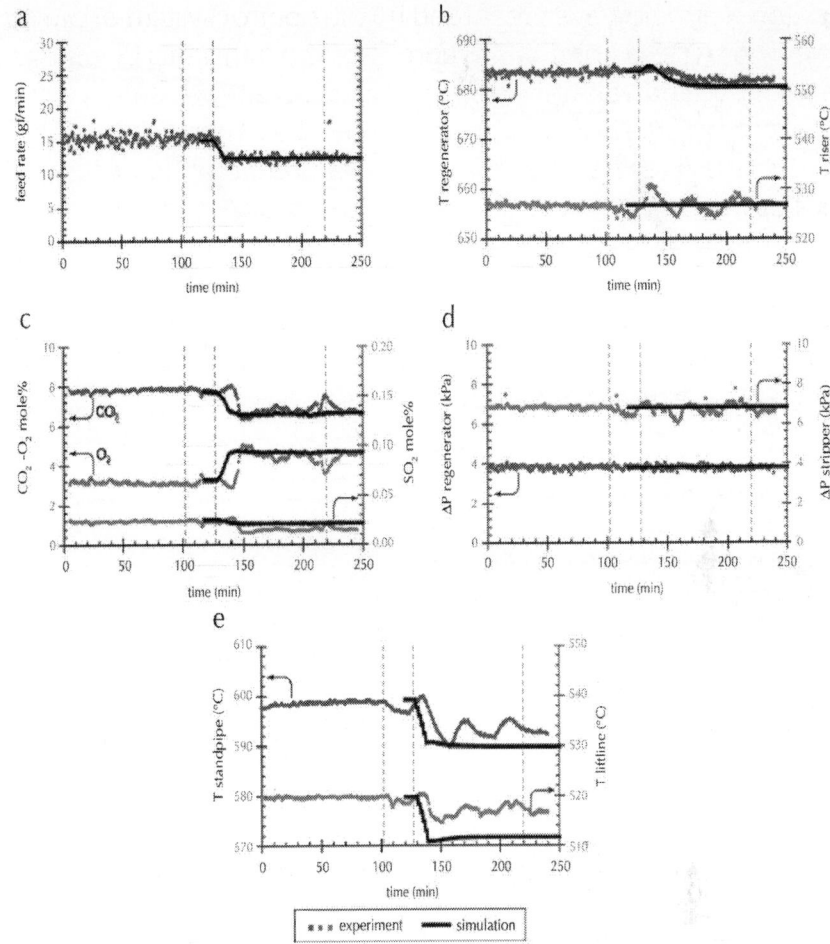

Figure 5: Closed-loop responses of pilot plant and simulator for a 15% decrease in feed rate.

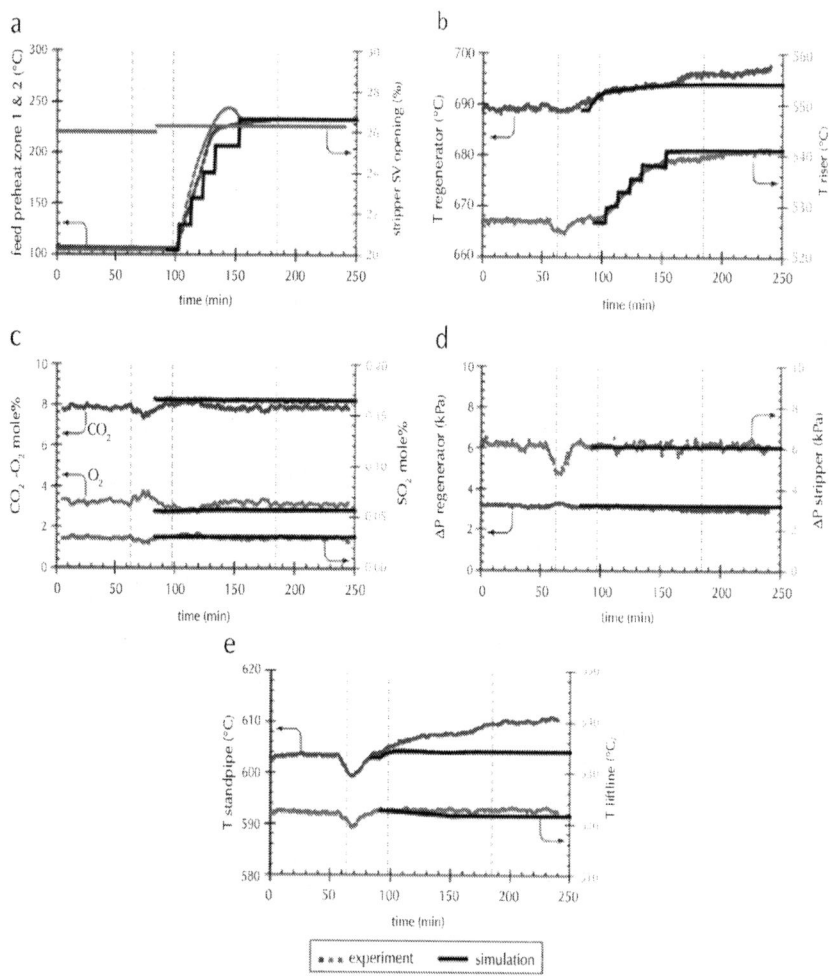

Figure 6: Open-loop responses of pilot plant and simulator for a 130% increase in feed preheat temperature.

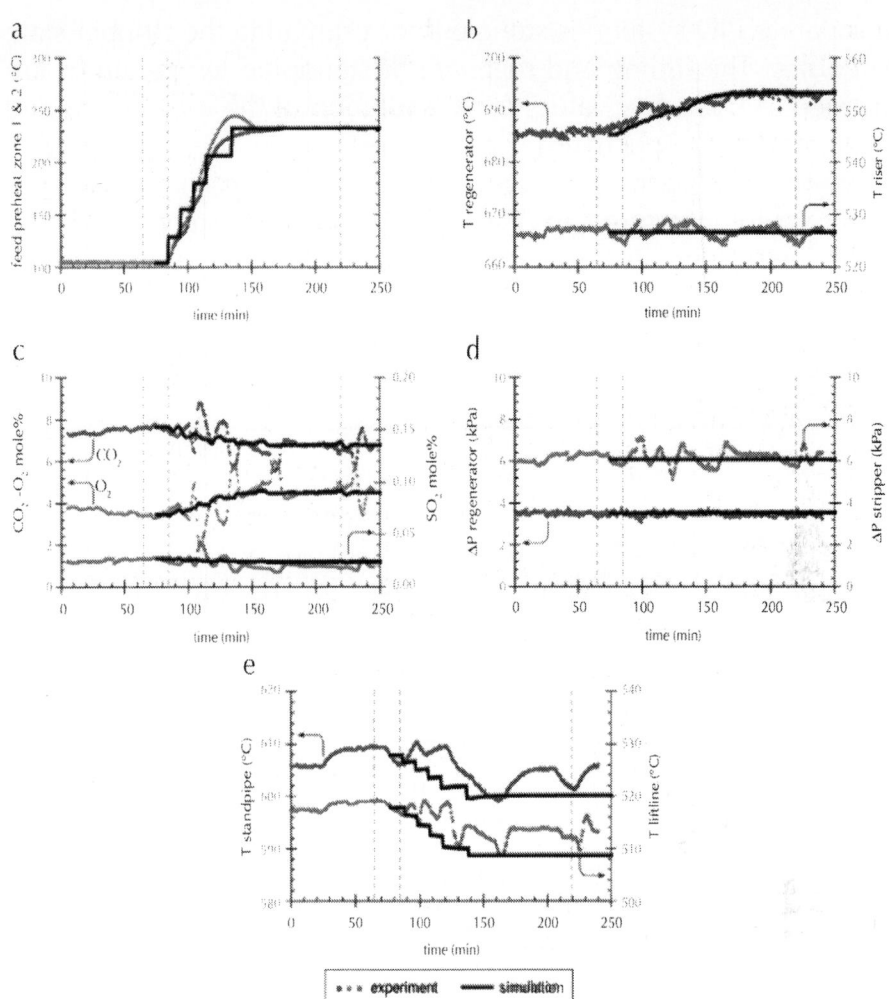

Figure 7: Closed-loop responses of pilot plant and simulator for a 130% increase in feed preheat temperature.

As the ulterior scope of this research work is the use of the simulator in an MPC structure for the pilot plant, it is useful to explain the sufficiency of the available pilot plant measurements. The regenerator temperature, pressure drop and flue gas composition are sufficient in order to estimate or provide an error feedback of the coke yield and the regenerator state variables. The stripper pressure drop (given that its temperature is controlled by a separate PID system) is sufficient for estimating the stripper state variables. The liftline and regenerator standpipe temperatures are needed as boundary values for the solution of the regenerator and riser models, respectively. Finally, the feed preheat temperature, catalyst circulation rate, regenerator standpipe exit temperature and riser temperature can provide an accurate inferential estimate of the instantaneous feed conversion that is not measured on line.

Feed Rate 15% Decrease—Open-Loop Operation

Experimental details: In the first experiment the regenerator standpipe slide valve was set to constant opening, equal to its average value of the previous 1 h steady-state operation. The time this action was taken is marked with the first vertical dotted line in the diagrams of Fig. 4. The average values introduced in such a nonlinear system produced momentary instabilities; hence a line-out period of 10 min was needed. When the unit reached steadiness (second vertical dotted line in diagrams of Fig. 4), a step change in the feed mass flow was imposed (from 15.1 to 12.4gmin^{-1}), which produced the transient in feed flow rate shown in Fig. 4(a). The feed flow rate that was entered to the simulator represented this transient by the consecutive step changes shown in Fig. 4(a). After a period of 20 min, the equivalent pattern of changes of Fig. 4(a) was manually imposed to the stripper standpipe slide valve. The specific oscillatory pattern was chosen to distinct the effect of catalyst circulation rate, without considerably altering the pressure balance of the unit. Under this pattern the ability of the simulator to depict the mixed effects of feed and catalyst rates in open-loop pilot plant operation was examined.

Effect of change in feed rate: The decrease in feed rate caused an instantaneous increase in riser temperature (Fig. 4(b)), because less feed consumed less heat for vaporization. Furthermore, the catalyst to oil ratio increased, since the catalyst circulation rate was the same (open-loop operation) and the feed rate lower. The latter resulted in higher conversion and coke yield (on feed basis) and this was observed and predicted, as shown in Table 4. However, the coke rate entering the regenerator decreased, because the feed rate was lower. This led to a decrease in the regenerator temperature (Fig. 4(b)). As shown in Fig. 4(c), the lower coke rate entering the regenerator resulted in a decrease in the CO_2 and SO_2 concentration and a parallel increase of the excess O_2. The decrease in regenerator temperature, that is, the input temperature of the standpipe, led to a smaller decrease in standpipe temperature (Fig. 4(e)), which again resulted in an even smaller decrease in riser temperature (Fig. 4(b)). The loop was continued for a period of 100 min (third vertical dotted line in diagrams of Fig. 4) until convergence.

Table 4: Steady-state results (experimental and predicted) of open- and closed-loop behavior of the pilot plant for a 15% decrease in feed rate

	Open-loop behavior		Closed-loop behavior	
Case examined	Steady state 1(a)	Steady state 1(b)	Steady state 2(a)	Steady state 2(b)
Feed rate Kgs^{-1}	25.19E−5	20.67E−5	25.64E−5	20.88E−5
Feed preheat (°C)	104.4	104.5	104.4	104.4

	Experimental vs. predicted operational variables				Experimental vs. predicted operational variables			
Catalyst to oil ratio	15.6	15.6	18.9	18.9	13.6	13.7	17.48	14.7
Riser temperature (°C)	526.8		535.5	534.7	526.8		526.8	
Reg. temperature (°C)	683.3	683.7	679.4	679.3	683.3	683.6	681.7	680.4
	Experimental vs. predicted yields				Experimental vs. predicted yields			
Conversion wt% on feed	66.8	65.8	69.6	70.7	64.7	63.7	68.4	65.8
Coke yield wt% on feed	5.78	5.77	6.63	6.66	5.31	5.32	6.49	5.65
Carbon wt% on reg. cat.	0.035	0.030	0.020	0.21	0.013	0.019	0.015	0.012

Indices (a) and (b) denote the initial and final (after the imposition of the step change) steady states. First and second columns in each steady state denote the experimental and the predicted variables, respectively. Single entries denote single inputs or set points.

Effect of change in catalyst flow: The comparison of Fig. 4(a)–(c) shows that the oscillation in the catalyst flow rate, produced by the changes in the stripper standpipe slide valve opening, showed a more pronounced effect on flue gas composition than on regenerator temperature. The reason of the latter is that the many factors involved in the regenerator heat balance (temperature and flow of catalyst and gas entering and exiting the regenerator, dense bed volume, heat loss) result in smoother transients of the regenerator temperature. The variations of the stripper slide valve opening led to implicit variation of the stripper level and thus in its pressure drop (Fig. 4(d)), without a parallel variation of the regenerator bed height. The reason for this is the much larger regenerator diameter. The change in regenerator temperature and the small variation of the catalyst flow rate resulted in the temperature profiles of liftline and standpipe shown in Fig. 4(e). The simulator predictions are in good agreement with the real dynamic behavior of the unit. The form of entering the step change in five representational consecutive steps has a negligible effect on the simulation of the unit dynamic responses. In open-loop operation the simulator can predict the dynamic behavior of the pilot plant in terms of temperature, yield of combustion reactions and pressure, which are the variables that can be measured on line.

Feed Rate 15% Decrease—Closed-Loop Operation

Experimental details: In this case the 15% decrease in feed flow rate was imposed (Fig. 5(a)) but for closed-loop operation of the unit. The regenerator standpipe slide valve was set to control the riser temperature and the stripper standpipe slide valve operated for constant stripper level. Accordingly, the routine of the simulator that adjusts the catalyst circulation for constant riser temperature was activated, while the catalyst rates at the entrance and the exit of the stripper were set equal to each other. The simulation results are of course quite different from the apparent dynamic behavior of the pilot plant that the actions of the controllers produce. Thus, the results of the simulator are examined in the sense of an "ideal efficiency controller". Nonetheless, the steady states and the general trends of the simulator and the pilot plant should be similar.

Effect of change in feed rate : The decrease in feed rate should have produced an increase in riser temperature, though the controller (or the equivalent adjustment routine) lowered the catalyst circulation rate to satisfy the heat balance of the riser for temperature equal to 526.8°C. The first regenerator response predicted by the simulator was a minor temperature increase, owed to the lower cold catalyst mass entering and hot catalyst mass exiting the regenerator. Thereafter, the regenerator temperature decreased due to the lower coke rate that decelerated the exothermic combustion reactions. As shown in Fig. 5(b), the phenomenon was quite different in the pilot plant. The regenerator standpipe slide valve controller was not efficient enough to balance the catalyst circulation rapidly, thus it produced an oscillation in the riser temperature for a period of 90 min. Moreover, the stripper bed height oscillated around its average value (Fig. 5(c)), whereas the regenerator bed height was again relatively constant, owing to its larger diameter. As shown in Fig. 5(c) and (d) the simulator results are close to the results of the pilot plant. The major difference between the simulator and the pilot plant is that the simulated responses are "faster" than those of the pilot plant. Nonetheless, the algorithmic adjustment of the catalyst circulation rate, used in the simulator, was expected to result in different dynamics than the PID controllers. The final steady state was achieved after 100 min in the unit, while a transient of only 40 min was predicted by the simulator. This is a typical example of how such a simulator can help towards the direction of optimal unit control if, for instance, the demand was for constant riser temperature and minimum transition time.

Feed Preheat Temperature 130% Increase—Open-Loop Operation

Experimental details : In this case the effect of increasing the feed preheat temperature from 104 to 232°C on the dynamic performance of the pilot plant in open-loop operation was explored. At the time marked with the first vertical dotted line in the diagrams of Fig. 6, the regenerator standpipe slide valve was set to constant opening. After a line-out period of 30 min (second vertical dotted line in diagrams of Fig. 6), the step change was imposed to the feed heater, which produced the transient in feed preheat temperature shown in Fig. 6(a). The long transient of preheat change was represented, for the simulator needs, by the profile of consecutive changes shown in Fig. 6(a). In this case the stripper standpipe slide valve was set to constant opening.

Effect of change in feed preheat: The increase in feed preheat temperature caused a rapid increase in the riser temperature, as the heat balance of the riser imposes (Fig. 6(b)). This increase in riser temperature led to higher feed conversion but not different coke yield (Table 5), since coke production is not significantly influenced by temperature (Bollas et al., 2002). The method of representing the transient of change in feed preheat with five consecutive steps is less accurate in this case. However, the general trends of the pilot plant and the simulator are similar (Fig. 6(b)). As shown in Fig. 6(b) and (e), there is a deviation of 2–5°C in the prediction of regenerator and standpipe temperature. This is not followed by a difference in the flue gas composition, which was predicted and observed relatively constant (Fig. 6(c)). Effects of wall heaters that operate for the establishment of pseudo-adiabatic conditions are responsible for this increase in the regenerator and standpipe temperature and were not used equivalently in the simulator. The pressure drops of the stripper and the regenerator were measured and predicted constant (Fig. 6(d)), because the catalyst circulation rate did not significantly change. As presented in Table 5, the accuracy in the prediction of the final steady state in this case is satisfactory.

Table 5: Steady-state results (experimental and predicted) of open- and closed-loop behavior of the pilot plant for a 130% increase in feed pre-heat temperature

	Open-loop behavior		Closed-loop behavior	
Case examined	Steady state 3(a)	Steady state 3(b)	Steady state 4(a)	Steady state 4(b)
Feed rate Kgs^{-1}	25.27E−5	25.25E−5	25.03E−5	25.26E−5
Feed preheat (°C)	104.4	232.4	104.4	232.3

	Experimental vs. predicted operational variables				Experimental vs. predicted operational variables			
Catalyst to oil ratio	14.8	14.7	14.6	14.5	13.5	13.5	12.9	11.21
Riser temperature (°C)	526.8		540.6	540.9	526.7		526.7	
Reg. temperature (°C)	688.1	688.8	696.8	693.9	685.4	685.3	694.4	693.3
	Experimental vs. predicted yields				Experimental vs. predicted yields			
Conversion wt% on feed	64.7	64.7	66.2	66.4	64.2	63.53	60.5	60.28
Coke yield wt% on feed	5.37	5.56	5.27	5.52	5.17	5.28	5.11	4.72
Carbon wt% on reg. cat.	0.030	0.037	0.025	0.033	0.010	0.023	0.015	0.013

Indices (a) and (b) denote the initial and final (after the imposition of the step change) steady states. First and second columns in each steady state denote the experimental and the predicted variables, respectively. Single entries denote single inputs or set points.

Feed Preheat Temperature 130% Increase— Closed-Loop Operation

Experimental details : The increase in feed preheat temperature from 104 to 232°C was applied but in closed-loop operation of the pilot plant (Fig. 7(a)). The slide valve controllers were set to automatic operation and the routine of the simulator that adjusts the catalyst circulation for constant riser temperature was activated. The imposed change produced the transient of Fig. 7(a), which again was represented by five consecutive step changes.

Effect of change in feed preheat: The inefficiency of the regenerator standpipe slide valve controller produced an oscillation in riser temperature (Fig. 7(b)), whereas the sluggish behavior of the stripper standpipe slide valve produced large oscillation in the stripper bed height and even a small oscillation in the regenerator bed height (Fig. 7(d)). The fluctuation of the catalyst stream entering the regenerator resulted in large oscillation in the flue gas composition (Fig. 7(c)). The imposed change led the pilot plant to a new steady state with lower catalyst circulation rate, that satisfied the riser heat balance, lower coke yield, because of the lower catalyst rate entering the riser, and higher regenerator temperature, because of the lower cold catalyst mass entering and hot catalyst mass exiting the regenerator, as shown in Table 5. The experimental results presented in Table 5 are of moderate accuracy, as in this case the final steady state was moderately established. The performance of the simulator is again "faster", yet in this case the difference between experiment and model is much more crucial. More robust control could definitely enhance the steadiness in the operation of the unit and would produce smoother profiles in critical unit variables as the riser temperature and the regenerator flue gas. However, the diverse process and operating conditions demand a more detuned performance of the controllers to stabilize the overall operation. The response of the simulator proposes that it could be possible with optimal model-based control to set the regenerator to absorb the increase in the feed preheat temperature and preserve the riser temperature constant at 526 °C with smooth unit performance.

CONCLUSIONS

A dynamic simulator of the FCC integrated riser–regenerator system was presented. The nonlinear dynamic and multivariable model was verified with a set of dynamic experiments carried out in the pilot plant of CPERI. The simulator performed satisfactorily in describing the complex responses of the unit to typical disturbances. It was evident through experiments with step changes in the riser input variables that the reactor has a very small contribution to the dynamic behavior of the integrated system, for which the regenerator and the stripper are mainly responsible. The excellent convergence between observed and predicted operating variables indicates the accurate formulation of the mass and energy balances. The results of both the simulator and the pilot plant are in excellent agreement with the experience of real-time operation of FCC units. The pilot plant of CPERI is mainly operated for performing catalyst benchmarking experiments at constant feed conversion and riser temperature. The ultimate scope of this research is to utilize the simulator in the development of a model-based control structure for the pilot plant, thus improving the process productivity. The effect of the model accuracy on the efficiency of a model-based controller is very important. It was demonstrated that the model presented has the ability to maintain good prediction properties over the range of operating conditions of interest. Therefore, a model-based control strategy is expected to allow for accurate targeting of prescribed operating conditions and the minimization of the number of experiments in catalyst evaluation tests.

ACKNOWLEDGMENT

Financial support by the Ministry of National Education and Religious Affairs (program 2.2.01 ARCHIMEDES I, EPEAEK II) and the General Secretariat of Research and Technology Hellas (program AKMON 01) is gratefully acknowledged.

APPENDIX A. HYDRODYNAMIC CORRELATIONS AND PRESSURE BALANCE

A.1. Simulation of Riser

The WHSV and the solids residence time (t_{ts}) were calculated by Eqs. (32) and (33), following the pilot riser geometry, that is divided into three regions:

$$\frac{WHSV}{3600}$$

$$= \frac{\dot{W}_{F:RS}}{\rho_p(V_{D:RS}(1-\varepsilon_{D:RS})+V_{C:RS}(1-\varepsilon_{C:RS})+V_{F:RS}(1-\varepsilon_{F:RS}))}, \quad (32)$$

$$t_{C:RS} = \frac{3600}{WHSV}\frac{\dot{W}_{F:RS}}{\dot{W}_{C:RS}}. \quad (33)$$

a. The mixing region at the riser bottom. The void fraction ($\varepsilon_{D:RS}$) and subsequently the catalyst inventory of this region were related to the superficial gas velocity by means of the empirical correlation of Richardson and Zaki (1954) (Eq. (34)), which substantiates for a dense regime in the bottom region of the pilot unit:

$$\varepsilon_{D:RS} = \left(\frac{u_{g:D:RS}}{u_{t:RS}} \right)^{1/z}.$$

(33)

b. The conical-shaped intermediate region. Because of the very small volume of the intermediate region (15% of total riser volume), a simple approximation of averaged (between top and bottom regions) hydrodynamic attributes was used (Pugsley and Berruti, 1996).

c. The fast fluidization region at the riser top, which was simulated under the following assumptions: (i) the flow is fully developed, thus its hydrodynamic features remain constant with height; (ii) the total volumetric yield of the reaction is flowing through the whole height of this region; (iii) the particle acceleration is considered to be negligible. These three assumptions can be accepted because the majority of the reactions taking place in the riser top region are secondary reactions of smaller molar expansion, hence Eq. (35) holds:

$$\varepsilon_{F:RS} = \frac{u_{g:F:RS}\rho_p A_{F:RS}}{y_{F:RS} \dot{W}_{C:RS} + u_{g:D:RS}\rho_p A_{F:RS}}.$$

(35)

In Eq. (35)$_{y_{F:RS}}$ is the average gas–solids slip factor for the top section of the riser, which was proven to play an important role in small-diameter riser reactors (Bollas et al., 2002). The correlation of Pugsley and Berruti (1996) was applied for the estimation of the gas–solids slip factor as shown in Eq. (36), where Fr_g and Fr_t are the Froudenumbers for the superficial gas velocity and solids terminal velocity, respectively:

$$y_{F:RS} = 1 + \frac{5.6}{Fr_{g:F:RS}^2} + 0.47 Fr_{t:F:RS}^{0.41}.$$

(36)

A detailed pressure gradient analysis is required for small-diameter risers (Bollas et al., 2002). For this analysis, all pressure gradients should be taken into account, and Eq. (37)is valid where Pfg is the gas–wall frictional pressure drop, Pfs is the solids–wall frictional pressure drop, Pacc is the pressure drop due to solids acceleration and the other terms represent the pressure drop due to solids and gas static head throughout the total riser height:

$$\Delta_{PRS} = \Delta_{Pfg:RS} + \Delta_{Pfs:RS} + \Delta_{Pacc:RS} + \varepsilon_{RS}\rho_{g:RS}gLRS + (1 - \varepsilon_{RS})\rho_{p}gLRS. \tag{37}$$

Simulation of Regenerator

For group A particles the emulsion gas superficial velocity is the gas velocity for zero net flow of solids, which equals the minimum bubbling velocity, plus (concurrent gas/solids flows) or minus (countercurrent gas/solids flow) the superficial solids velocity in the emulsion phase:

$$u_{ge} = u_{mb} \pm u_{se}. \tag{38}$$

For the evaluation of the minimum fluidization velocity the equation of Wen and Yu (Kunii and Levenspiel, 1977) is applied. For group A particles the minimum bubbling velocity, u_{mb}, is evaluated by the correlation of Abrahamsen and Geldart (1980), which considers the effect of catalyst fines, f, on u_{mb}:

$$\frac{u_{mb}}{u_{mf}} = \frac{2300\rho_{ge}^{0.126}\mu_{ge}^{0.523}\exp(0.716f)}{d_{p}^{0.8}g^{0.934}(\rho_{p} - \rho_{ge})^{0.934}}. \tag{39}$$

The superficial gas velocity in the dense zone is then obtained by

$$u_{g}:R_{G} = u_{gb} + u_{ge}. \tag{40}$$

The fraction of the bubbles in the dense zone is

$$f_b = \frac{u_{gb}}{v_b}. \tag{41}$$

The absolute bubble rise velocity v_b is calculated as a function the isolated bubble rise velocity:

$$\upsilon_b = 0.711(g_{db})^{0.5} + u_{g:RG} - u_{ge}. \tag{42}$$

The bubble diameter is estimated by the Wen–Mori correlation (Kunii and Levenspiel, 1977):

$$\frac{d_b^{(l_D=1)} - d_b}{d_b^{(l_D=1)} - d_b^{(l_D=0)}} = \exp\left(-\frac{0.3L_{D:RG}l_D}{D_{D:RG}}\right). \tag{43}$$

The initial bubble diameter and the maximum bubble diameter are estimated by Eqs.(26) and (27), respectively:

$$d_b^{(l_D=0)} = \frac{1.38}{g^{0.2}}\left(\frac{1}{1000}(u_{g:RG}^{(l_D=0)} - u_{mb})\right)^{0.4}, \tag{44}$$

$$d_b^{(l_D=1)} = \min\left[0.652(A_{D:RG}(u_{g:RG}^{(l_D=1)} - u_{mb}))^{0.4}, 2\frac{(u_t^{(2.7d_p)})^2}{g}\right]. \tag{45}$$

The emulsion to freeboard elutriation rate K_i^* of a fraction of particles with average diameter dpi is evaluated by the Zenz and Weil correlation (Geldart, 1985). The total entrainment rate K_i^* is then obtained by adding the rates of each respective fraction of particles. The catalyst density in the freeboard is a function of the gas–solids slip velocity, which is calculated on the basis of the correlation of Patience et al. (1992) as shown in

$$u_{sf} = \frac{u_{gf}}{1 + 5.6/Fr_{gf} + 0.47Fr_t^{0.41}}. \tag{46}$$

The freeboard voidage is then calculated by

$$\varepsilon_f = 1 - \frac{K_t^*}{\rho_p u_{sf}}. \tag{47}$$

The pressure drop throughout the regenerator is calculated from the solids static head as shown in

$$\Delta_{PRG} = {}_{\rho p}(1 - {}_{\varepsilon e})f_{egLD:RG} + {}_{\rho p}(1 - {}_{\varepsilon f})g_{LF:RG}. \tag{48}$$

A.3. Simulation of Stripper and Slide Valves

The pressure drop throughout the stripper is calculated from the solids static head as shown in

$$\Delta_{PST} = {}_{\rho p}(1 - {}_{\varepsilon mf})g_{LD:ST}. \tag{49}$$

The catalyst circulation rate at the entrance and exit of the regenerator was correlated with the slide valve opening of the stripper and regenerator standpipes and the pressure drop by (Judd and Dixon, 1978)

$$\dot{W}_C = k_{SV} \left(\frac{A_0^2 A_{t:SV}^2}{A_0^2 - A_{t:SV}^2} \right)^{0.5} (2\rho_p(1 - \varepsilon)\Delta P_{SV})^{0.5}. \tag{50}$$

REFERENCES

1. Abrahamsen, A.R., Geldart, D., 1980. Behaviour of gas-fluidized beds of fine powders, part I. Homogeneous expansion. Powder Technology 26 (1), 35–46.

2. Ali, H., Rohani, S., 1997. Dynamic modeling and simulation of a riser-type fluid catalytic cracking unit. Chemical Engineering and Technology 20 (2), 118–130.

3. Ali, H., et al., 1997. Modelling and control of a riser-type fluid catalytic cracking (FCC) unit. Chemical Engineering Research and Design 75 (A4), 401–412.

4. Arbel, A., et al., 1995a. Dynamics and control of fluidized catalytic crackers. 1. Modeling of the current generation of FCCs. Industrial and Engineering Chemistry Research 34 (4), 1228–1243.

5. Arbel, A., et al., 1995b. Dynamics and control of fluidized catalytic crackers. 2. Multiple steady-states and instabilities.

Industrial and Engineering Chemistry Research 34 (9), 3014–3026.

6. Arthur, J.R., 1951. Reactions between carbon and oxygen. Transactions of the Faraday Society 47, 164–178.

7. Blanding, F.H., 1953. Reaction rates in the catalytic cracking of petroleum. Industrial and Engineering Chemistry 45 (6), 1186–1197.

8. Bollas, G.M., et al., 2002. Modeling small-diameter FCC riser reactors. A hydrodynamic and kinetic approach. Industrial and Engineering Chemistry Research 41 (22), 5410–5419.

9. Bollas, G.M., et al., 2004. Bulk molecular characterization approach for the simulation of FCC feedstocks. Industrial and Engineering Chemistry Research 43 (13), 3270–3281.

10. Cristea, M.V., et al., 2003. Simulation and model predictive control of a UOP fluid catalytic cracking unit. Chemical Engineering and Processing 42 (2), 67–91.

11. Davidson, J.F., et al., 1985. Fluidization. Academic Press, London. Elnashaie, S., Elshishini, S.S., 1993. Digital-simulation of industrial fluid catalytic cracking units. 4. Dynamic behavior. Chemical Engineering Science 48 (3), 567–583.

12. Elnashaie, S., et al., 1995. Digital-simulation of industrial fluid catalytic cracking units. 5. Static and dynamic bifurcation. Chemical Engineering Science 50 (10), 1635–1644.

13. Faltsi-Saravelou, O., Vasalos, I.A., 1991. Fbsim—a model for fluidized-bed simulation. 1. Dynamic modeling of an adiabatic reacting system of small gas-fluidized particles. Computers and Chemical Engineering 15 (9), 639–646.

14. Faltsi-Saravelou, O., et al., 1991. Fbsim—a model for fluidized-bed simulation. 2. Simulation of an industrial fluidized catalytic cracking regenerator. Computers and Chemical Engineering 15 (9), 647–656.

15. Ford, W.D., et al., 1977. Operating cat crackers for maximum profit. Chemical Engineering Progress 73 (4), 92–96.

16. Geldart, D., 1973. Types of gas fluidization. Powder Technology 7 (5), 285–292.

17. Geldart, D., 1985. Elutriation. In: Davidson, J.F. et al. (Eds.), Fluidization. Academic Press, London, p. 383.

18. Han, I.S., Chung, C.B., 2001a. Dynamic modeling and simulation of a fluidized catalytic cracking process. Part I: Process modeling. Chemical Engineering Science 56 (5), 1951–1971.

19. Han, I.S., Chung, C.B., 2001b. Dynamic modeling and simulation of a fluidized catalytic cracking process. Part II: Property estimation and simulation. Chemical Engineering Science 56 (5), 1973–1990.

20. Han, I.S., et al., 2004. Modeling and optimization of a fluidized catalytic cracking process under full and partial combustion modes. Chemical Engineering and Processing 43 (8), 1063–1084.

21. Hernandez-Barajas, J.R., et al., 2006. Multiplicity of steady states in FCC units: effect of operating conditions. Fuel 85 (5–6), 849–859.

22. Howard, J.B., et al., 1973. Kinetics of Carbon Monoxide Oxidation in Post Flame Gases. In: Fourteenth Symposium (International) on Combustion, Pittsburgh, The Combustion Institute, 975–986.

23. Jacob, S.M., et al., 1976. Lumping and reaction scheme for catalytic cracking. A.I.Ch.E. Journal 22 (4), 701–713.

24. Judd, M.R., Dixon, P.D., 1978. Flow of fine, dense solids down a vertical standpipe. A.I.Ch.E. Symposium Series 74 (176), 38–44.

25. Kesler, M.G., Lee, B.I., 1976. Improve prediction of enthalpy of fractions. Hydrocarbon Processing 55 (3), 153–158.

26. Kunii, D., Levenspiel, O., 1977. Fluidization Engineering. Robert E. Krieger Publishing Company Inc, Florida.

27. Lee, E., Groves Jr., F.R., 1985. Mathematical model of the fluidized bed catalytic cracking plant. Transactions of the Society for Computer Simulation International 2 (3), 219–236.

28. Lopez-Isunza, F., 1992. Dynamic modeling of an industrial fluid catalytic cracking unit. Computers and Chemical Engineering 16, S139–S148.

29. McFarlane, R.C., et al., 1993. Dynamic simulator for a model-IV fluid catalytic cracking unit. Computers and Chemical Engineering 17 (3), 275–300.

30. Morley, K., De Lasa, H.I., 1987. On the determination of kinetic parameters for the regeneration of cracking catalyst. Canadian Journal of Chemical Engineering 65 (5), 773–777.

31. Patience, G.S., et al., 1992. Scaling considerations for circulating fluidized bed risers. Powder Technology 72 (1), 31–37.

32. Pugsley, T.S., Berruti, F., 1996. A predictive hydrodynamic model for circulating fluidized bed risers. Powder Technology 89 (1), 57–69.

33. Richardson, J.F., Zaki, W.N., 1954. Sedimentation and fluidization. I. Transactions of the Institution of Chemical Engineers 32, 35–53.

34. Secchi, A.R., et al., 2001. A dynamic model for a FCC UOP stacked converter unit. Computers and Chemical Engineering 25 (4–6), 851–858.

35. Tone, S., et al., 1972. Kinetics of oxidation of coke on silica-alumina catalysts.

36. Bulletin of Japan Petroleum Institute 14 (1), 76–80. Vasalos, I.A., et al., 1996. Design construction and experimental results of a

37. CFB FCC pilot plant. In: Kwauk, V.M., Li, J. (Eds.), Circulating Fluidized Bed Technology V. Science Press, Beijing, China.

38. Voorhies, A., 1945. Carbon formation on catalytic cracking. Industrial and Engineering Chemistry 37 (4), 318–322.

39. Wang, G.-X., 1986. Kinetics of combustion of carbon and hydrogen in carbonaceous deposits on zeolite-type cracking catalysts. Industrial & Engineering Chemistry, Process Design and Development 25 (3), 626–630.

40. Weekman, V.M., Nace, D.M., 1970. Kinetics of catalytic cracking selectivity in fixed, moving, and fluid bed reactors. A.I.Ch.E. Journal 16 (3), 397–404.

41. Wollaston, E.G., et al., 1975. What influences cat cracking. Hydrocarbon rocessing 54 (9), 93–100.

Citations

CHAPTER 1

H. C. Alvarez-Castro, E. M. Matos, M. Mori, W. Martignoni, and R. Ocone, "Analysis of Process Variables via CFD to Evaluate the Performance of a FCC Riser," International Journal of Chemical Engineering, vol. 2015, Article ID 259603, 13 pages, 2015. doi:10.1155/2015/259603.

CHAPTER 2

Liang Yang, Zhongli Ji, Qiaoqi Xu, and Hao Li, "Performance Assessment of Sintered Metal Fiber Filters in Fluid Catalytic Cracking

Unit," International Journal of Chemical Engineering, vol. 2014, Article ID 371853, 10 pages, 2014. doi:10.1155/2014/371853.

CHAPTER 3

Gabriela C. Lopes, Leonardo M. Rosa, Milton Mori, José R. Nunhez, and Waldir P. Martignoni, "CFD Study of Industrial FCC Risers: The Effect of Outlet Configurations on Hydrodynamics and Reactions," International Journal of Chemical Engineering, vol. 2012, Article ID 193639, 16 pages, 2012. doi:10.1155/2012/193639.

CHAPTER 4

P. Dasila, I. Choudhury, D. Saraf, S. Chopra and A. Dalai, "Parametric Sensitivity Studies in a Commercial FCC Unit," Advances in Chemical Engineering and Science, Vol. 2 No. 1, 2012, pp. 136-149. doi:10.4236/aces.2012.21017.

CHAPTER 5

Zheng, J., Lang, X. and Wang, C. (2014) Effect of Preparation Method on Catalytic Properties of Double Perovskite Oxides LaSrFeMo0.9Co0.1O6 for Methane Combustion. Advances in Chemical Engineering and Science, 4, 367-373. doi: 10.4236/aces.2014.43040.

CHAPTER 6

Deng, Z. , Bai, F. , Xing, Y. , Xing, N. and Xu, L. (2013) Reaction in situ found in the synthesis of a series of lanthanide sulfate complexes and investigation on their structure, spectra and catalytic activity. Open Journal of Inorganic Chemistry, 3, 76-99. doi: 10.4236/ojic.2013.34011.

CHAPTER 7

G.M. Bollas, I.A. Vasalos, A.A. Lappas, D.K. Iatridis, S.S. Voutetakis, S.A. Papadopoulou, Integrated FCC riser—regenerator dynamics studied in a fluid catalytic cracking pilot plant, Chemical Engineering Science, Volume 62, Issue 7, April 2007, Pages 1887-1904, ISSN 0009-2509, http://dx.doi.org/10.1016/j.ces.2006.12.042.

Index